Connecting with Ambivalent Heritage

UCL Critical Cultural Heritage Series

Series Editor: Beverley Butler

The aim of this series is to ask new questions about what heritage is and does, and why it is important. Rejecting the idea that there is any one static model, it seeks out as-yet-unconceptualized notions of heritage based on grounded research contexts and cutting-edge methodologies that are global in reach and traverse disciplinary boundaries to draw on diverse perspectives notably, though not limited to, archaeology and museum studies, anthropology, ethnography, political studies, postcolonial studies, cultural memory studies, health humanities and environmental humanities. Published in association with the UCL Institute of Archaeology, the series welcomes submissions for authored and edited volumes from scholars and practitioners worldwide in pursuit of its goals to embrace a plurality of perspectives, increase diversity and representation in the discipline, and exemplify a proactive, responsive and just future for critical cultural heritage studies.

A full list of volumes published in the series before its move to Bloomsbury is available online at https://www.ucl.ac.uk/archaeology/about-us/institute-archaeology-publications.

Also available in the series

Narrating Heritage: Rights, Abuses and Cultural Resistance by Veysel Apaydin

Connecting with Ambivalent Heritage

Creative Uses of Postindustrial Spaces

Edited by Tiina Äikäs and Tuuli Matila

BLOOMSBURY ACADEMIC
LONDON • NEW YORK • OXFORD • NEW DELHI • SYDNEY

BLOOMSBURY ACADEMIC
Bloomsbury Publishing Plc
50 Bedford Square, London, WC1B 3DP, UK
1385 Broadway, New York, NY 10018, USA
29 Earlsfort Terrace, Dublin 2, Ireland

BLOOMSBURY, BLOOMSBURY ACADEMIC and the Diana logo are trademarks of Bloomsbury Publishing Plc

First published in Great Britain 2024

Cover design: Terry Woodley
Cover image: An abandoned brick factory in southern Finland © Photo by Oula Seitsonen

A catalogue record for this book is available from the British Library.

Library of Congress Cataloging-in-Publication Data
Names: Äikäs, Tiina, editor. | Matila, Tuuli, editor.
Title: Connecting with ambivalent heritage : creative uses of postindustrial spaces / [edited by] Tiina Äikäs and Tuuli Matila.
Description: London ; New York : Bloomsbury Academic, 2024. | Series: UCL critical cultural heritage series | This book is based on a workshop called "Connecting with the Ambivalent Heritage: Creative Ways to (Re)use Industrial Spaces" organised by the Smokestack memories 17 project held in Oulu in October 2022. | Includes bibliographical references and index.
Identifiers: LCCN 2024003279 (print) | LCCN 2024003280 (ebook) | ISBN 9781350426740 (hardback) | ISBN 9781350426788 (paperback) | ISBN 9781350426757 (pdf) | ISBN 9781350426764 (ebook)
Subjects: LCSH: Historic industrial sites–Europe, Northern. | Cultural property–Social aspects.
Classification: LCC NA9053.I53 C66 2024 (print) | LCC NA9053.I53 (ebook) | DDC 363.6/90948--dc23/eng/20240316
LC record available at https://lccn.loc.gov/2024003279
LC ebook record available at https://lccn.loc.gov/2024003280

ISBN: HB: 978-1-3504-2674-0
PB: 978-1-3504-2678-8
ePDF: 978-1-3504-2675-7
eBook: 978-1-3504-2676-4

Series: UCL Critical Cultural Heritage Series

Typeset by RefineCatch Limited, Bungay, Suffolk

To find out more about our authors and books visit www.bloomsbury.com and sign up for our newsletters.

Contents

Illustrations

Tables

Contributors

Tiina Äikäs is University Researcher in Archaeology at the University of Oulu, Finland. She is the Principal Investigator (PI) of the three-year project 'Smokestack Memories'. Her research interests focus on industrial heritage as well as contemporary archaeology, especially contemporary use of sacred places. Her latest book is *Shamanic Materialities in Nordic Climates* (2023) co-authored with Trude Fonneland.

Daniel Brandt is Associate Professor in Human Geography at Dalarna University in the Department of Culture and Society. He is a member of the Center for Tourism and Leisure Research (CeTLer), and his research focus is on the economic and spatial dimension of tourism. He has previously published articles on interpretation of heritage sites and development of digital information systems. His most recent article is on building a spatial decision support system for tourism and infrastructure planning (2022).

Minni Haanpää (PhD) works as Senior Lecturer in Tourism Studies at the University of Lapland, Faculty of Social Sciences, Multidimensional Tourism Institute (MTI). Her research interests include tourism and event experiences, affect, value co-creation and ethnographic methodologies, especially videography and autoethnography. Her current research projects explore the possibilities of tourism as social change in the Nordic urban context and technology-mediated tourism experiences. Her work has been published in *Qualitative Market Research*, *Journal of Marketing Management*, *Journal of Business Ethics*, *Tourism Geographies*, *Tourism Recreation Research*, *Society and Leisure* and in edited, peer-reviewed books.

Maria Hakkarainen (PhD) works as Senior Lecturer in Tourism Studies at the University of Lapland, Faculty of Social Sciences, MTI. Her research interests focus on local involvement in tourism, tourism ecosystems, sharing economy in tourism, participatory research methods and project development. She does research, teaches and trains within her areas of interest, and has also authored and co-authored publications in tourism journals and book chapters.

Hakkarainen has been involved in several national and regional level tourism development processes and projects in Finland. Currently she works as a Principal Investigator in the 'Travelers' Hospitality and Mobility Ecosystem, FIT ME!' project.

Anu Harju-Myllyaho is currently employed at Lapland University of Applied Sciences, where she holds the position of Senior Specialist in Research and Development. Harju-Myllyaho's research is primarily focused on inclusive tourism, tourism workforce development and tourism foresight. Recently she has edited a book *Inclusive Tourism Futures* (2021). In her doctoral dissertation, Harju-Myllyaho focuses on futures studies methods and inclusive tourism.

Vesa-Pekka Herva is Professor of Archaeology at the University of Oulu, Finland. He has studied various aspects of material culture, human–environment relations and cosmology in the high North of Europe from the Neolithic to the present day. He is the PI of the research project 'Extractive Industries as Engagement with the Extraordinary Subterranean: Culture, Heritage and Impact of Resource Extraction in Northernmost Europe' (Academy of Finland, 2021–5). He is the author, with Antti Lahelma, of *Northern Archaeologies and Cosmologies: A Relational View* (2020).

Marjo Anneli Juola is a PhD researcher in archaeology at the University of Oulu, Finland. She has been a part of the 'Smokestack Memories' project, where her research has been dealing with ruination and multispecies interactions in the postindustrial landscape. She is currently doing her research in a transdisciplinary research project 'Archiving the Planet: Material Infrastructures and Ecologies of Storage in the High North' (SEEDARC).

Eerika Koskinen-Koivisto is a Senior Researcher in Ethnology and Cultural Anthropology at the University of Jyväskylä and PI of the SENSOMEMO project (20200–2024 Research Council of Finland). Her research interests include material heritage, sensory memory and ethnographic methods. Her publications include *Her Own Worth* (Finnish Literature Society 2014) and the co-edited volumes *Reconstructing Homes: Affective Materiality and Atmospheres of Belonging* (2024), *Challenges and Solutions in Ethnographic Research: Ethnography with a Twist* (2020), *Routledge Handbook of Memory and Place* (2019), and *Transnational Death* (Finnish Literature Society 2019).

Teresa Komu is a postdoctoral researcher in the Arctic Centre, the University of Lapland and the University of Oulu, Finland, where she is a team member of the project 'Extractive Industries as Engagement with the Extraordinary Subterranean: Culture, Heritage and Impact of Resource Extraction in Northernmost Europe' (Academy of Finland, 2021–5). She has a PhD in cultural anthropology, with a thesis on the co-existence of reindeer herding, extractive industries and nature-based tourism in northern Fennoscandia. Her research interests currently circle around the coexistence of competing livelihoods, northern human–environment relations and the anthropology of the good.

Pauliina Latvala-Harvilahti, Adjunct Professor of Cultural Heritage Studies, is Senior Researcher in the Finland Futures Research Centre at the University of Turku, Finland. Her research interests focus on living heritage, international and national heritage politics, heritage communities, cultural policy, human–nature relationship, future agency and cultural sustainability transformation. She has recently published on environmental art reception, artivism and biocultural heritage.

Francisco Martínez is a researcher at the WasteMatters group of Tampere University. He is an anthropologist dealing with contemporary issues of material culture through ethnographic experiments. In 2018, he was awarded the Early Career Prize of the European Association of Social Anthropologists. Currently, he works at Tampere University and convenes the Collaboratory for Ethnographic Experimentation (EASA Network). Francisco has published several books, including *Ethnographic Experiments with Artists, Designers and Boundary Objects* (2021), *Remains of the Soviet Past in Estonia* (2018) and *Repair, Brokenness, Breakthrough* (2019). He has also curated different exhibitions.

Gabriel Moshenska is Associate Professor in Public Archaeology at University College, London. His research interests range wildly from gas masks and ghost stories to internet memes and the contested reception of Milton's theological writings in the nineteenth century. He is a collaborative researcher on the project 'Extractive Industries as Engagement with the Extraordinary Subterranean: Culture, Heritage and Impact of Resource Extraction in Northernmost Europe' (Academy of Finland, 2021–5). He is the author of more than 100 publications across several disciplines, including *Material Cultures of Childhood in Second World War Britain* (2019) and *Key Concepts in Public Archaeology* (2017).

Maija Mäki is Lecturer in Museum Studies at the University of Turku, Finland. Her research interests focus on heritage and futures, which often deal with the complex relations between citizens and authorities. Her recent publications address issues pertaining to urban heritage, for example, can urban heritage serve as the anchor for an uncertain future, discussing particularly the City of Turku and the Covid-19 crisis (2022, co-author Visa Immonen). She is currently working as editor-in-chief for the international journal of the Association of Finnish Ethnologists, *Ethnologia Fennica*.

Tuuli Matila is a postdoctoral researcher in the Department of Archaeology in the University of Oulu. Her research focuses on issues surrounding visuality, heritage and contemporary archaeology. Her dissertation *Seeing the War Through a Finnish Lens: Representation and Affect in the World War Two Photographic Heritage* (2022) was the first in Finland that examined visuality in a museum setting.

Risto Nurmi is an archaeologist and researcher in the project 'Extractive Industries as Engagement with the Extraordinary Subterranean: Culture, Heritage and Impact of Resource Extraction in Northernmost Europe' (Academy of Finland, 2021–5) at the University of Oulu. He gained a PhD in archaeology in 2011 with his thesis on urban archaeology. He specializes in historical archaeology and material culture studies in northern Europe and has published in many books and journals, including *International Journal of Historical Archaeology* and *Journal of Social Archaeology*.

Veera Ojala is a PhD researcher in cultural heritage at the University of Turku, Department of Humanities, in Finland. Her research revolves around a multidisciplinary research orientation that combines nuclear power and cultural heritage research under the heading 'Nuclear Cultural Heritage'. With her ongoing investigation of the visual culture of the Chernobyl Exclusion Zone (CEZ), she has gained insights into the public's role in the interpretation and archive processes of nuclear cultural heritage. Her most recent publication elaborates on the visual storytelling of visitors to the CEZ through three different visitor profiles (2022).

Hilary Orange is Lecturer in Heritage at Swansea University in the Department of History, Heritage and Classics. Her research interests focus on the contemporary archaeology of popular music, deindustrialization and industrial heritage. She has published on industrial landscape in Germany, the UK and Japan. In 2019, she co-edited the *Routledge Handbook of Memory and Place*.

Tina Paphitis is Associate Professor of Cultural Studies at the University of Bergen. She is a collaborative researcher on the project 'Extractive Industries as Engagement with the Extraordinary Subterranean: Culture, Heritage and Impact of Resource Extraction in Northernmost Europe' (Academy of Finland, 2021–5). Her research interests include legends and landscapes in Britain and the Nordic region, environmental/ecocritical folklore, folklore and archaeology in fantasy and horror literature, and folklore in critical heritage studies.

Albina Pashkevich is Associate Professor in Tourism Studies, Senior Lecturer in Human Geography at the School of Culture and Society, Dalarna University, Sweden. Her current research interests include investigations concerning tourism development of territories on the fringe and peripheral regions, as well as a study of institutional settings allowing for the alternative sustainable development of local communities, based on the cultural heritage of former mining.

Helena Ruotsala is Professor Emerita of European Ethnology at the University of Turku, Finland. She has studied environmental changes and reindeer herding from the standpoint of ecological and human–animal relations. She has been doing fieldwork in a rural Mari village in central Russia, where she is focusing on everyday life and identity issues in multicultural spaces. Her recent studies have focused on transnationalism and everyday life in the twin border cities of Tornio, Finland and Haparanda, Sweden.

Oula Seitsonen is an archaeologist and geographer and currently works as a postdoctoral researcher at the University of Oulu, Finland. His interests cover a wide temporal and thematic range from prehistoric pastoralism to contemporary archaeology. He is involved in various projects including 'Extractive Industries as Engagement with the Extraordinary Subterranean: Culture, Heritage and Impact of Resource Extraction in Northernmost Europe' (Academy of Finland, 2021–5). He is the author of *Archaeologies of Hitler's Arctic War: Heritage of the Second World War German Military Presence in Finnish Lapland* (2021).

Alexa D. Spiwak is a doctoral candidate at the University of Oslo in the Department of Archaeology, Conservation and History. Her research interests include industrial heritage, landscape studies, heritage policy and management, and natural–cultural entanglements.

Preface

Recent years have witnessed the demolition of old factories and the dismantling of smokestacks. Newspaper comment sections depict the annoyance and sadness that residents have expressed at losing parts of their local history. On the contrary though, some stress the ugliness, unusable and dangerous nature of these structures. Similarly in heritage discourse, some remains of the industrial past are given UNESCO status, whereas others are left to ruination and in their ruined form these abandoned structures might attract uses generally not seen as proper for heritage sites by the heritage authorities. We have explored these sometimes contrasting uses in our current project 'Smokestack Memories' (2021–3), where we approach the memories, values and re-use of factory smokestacks, as well as surrounding industrial remains, from a heritage perspective. The project has included crowdsourcing industrial heritage sites via an interactive online map and collecting photos and memories; interviews of different users of the sites have not been limited to former workers; and art projects have involved locals. The artist in our team, Suvi Nurmi, has used material from our transcribed interviews, and by printing out lines and words from them has let the local residents create poems from them.

This book is based on a workshop called 'Connecting with the Ambivalent Heritage: Creative Ways to (Re)use Industrial Spaces' organized by the 'Smokestack Memories' project held in Oulu in October 2022. The book broadens the topic and considers the multiple afterlives of industrial structures. The poem introduced in this volume was created by one of the workshop participants from the initiative of Suvi Nurmi using words from the e-mails exchanged between the seminar's participants and their paper abstracts. The workshop raised many novel ways to think about industrial heritage and its meaning to people, and how these might, as opposed to conservation, be understood through different usages and interactions with the spaces. It also highlighted the agency of the people involved and the wide variety of activities that are carried out in these sites, which are not always 'officially' considered heritage. This volume, which brings forth the ambivalence of the sites, could be seen as enabling these different interactions and re-uses.

Acknowledgements

We wish to thank the Kone Foundation for funding the project 'Smokestack Memories' (2021–3), which enabled us to work with the heritage of factory smokestacks. We would also like to thank the Biodiverse Anthropocenes programme at the University of Oulu for its support for organizing the 'Connecting with the Ambivalent Heritage: Creative Ways to (Re)use Industrial Spaces' workshop and for publishing this volume.

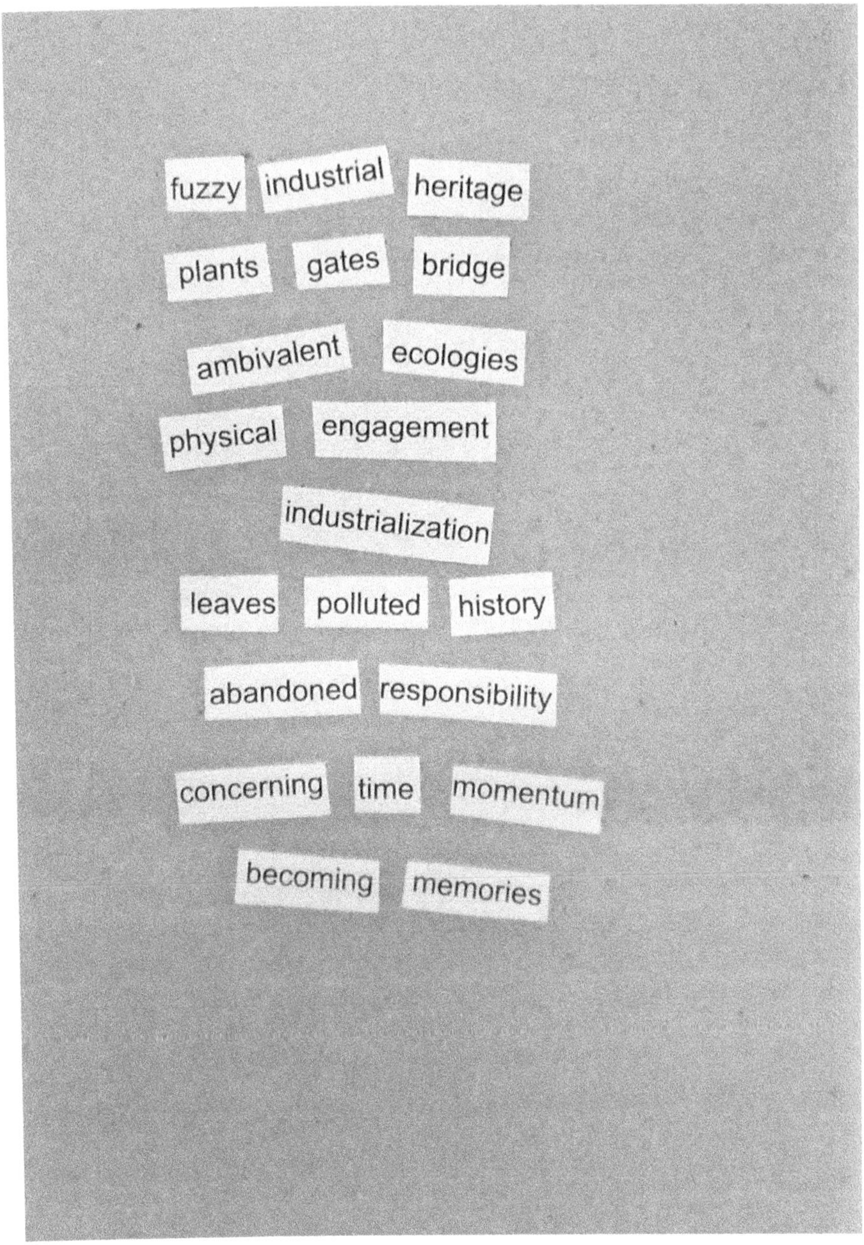
fuzzy industrial heritage
plants gates bridge
ambivalent ecologies
physical engagement
industrialization
leaves polluted history
abandoned responsibility
concerning time momentum
becoming memories

INTRODUCTION

Connecting with Ambivalent Heritage: Creative Uses of Postindustrial Spaces

Eerika Koskinen-Koivisto, Tiina Äikäs and Pauliina Latvala-Harvilahti

Ambivalent heritage: Exploring the uses of postindustrial spaces

In most dystopias, our planet and human life, as we know it, goes through a radical change or a stop. More realistic scenarios reflect life in transformed and ruined environments, in which humans continue to adjust and find ways to cohabitate them with materials and other species, and to use them for future livelihoods and dwelling. The future landscapes contain layers of past human activities, including fossil-based industry, but they gain new meanings. Industrial heritage is an interesting example of transforming/transient and dynamic heritage: as global industries develop and constantly move to a place where the circumstances are most profitable and raw materials abundant, new industrial sites and communities emerge, live for some while and then fall. The physical sites often remain, but are re-used or abandoned. The industrial plants, factories or mines, for instance, may have affected the local environment, leaving visible marks and altered biospheres in the landscape. The communities created around industries disappear but may continue to be remembered through sites of memory. The cycles of industrialization and deindustrialization also leave behind visible landmarks, such as smokestacks, which become symbols of the past that can raise mixed emotions of nostalgia. All this gives industrial heritage a specific ambivalent essence also in the processes of memorialization.

A dictionary definition for ambivalence sees it as 'simultaneous and contradictory attitudes or feelings (such as attraction and repulsion) toward an object, person, or action' (https://www.merriam-webster.com/dictionary/ambivalence). By using the concept ambivalent heritage, we put the emphasis on different ways to approach, use and value it, and the sometimes conflicting and confusing attitudes on how to use and preserve these sites. Hence the focus is not on the administrative status of the sites but on the experiences and emotions of users and visitors.

In ambivalent heritage, a range of feelings and affective meanings are present at the same time. Anthropologist Emily Donaldson (2018) has shown, in the indigenous context of the Marquesas Islands, how places of pain and discomfort have contributed to shared identity and supportive place-making. But ambivalence can also lead to the discard and abandonment of sites. This is the case in the postcolonial reality of the imperialist Park Street cemetery in Calcutta, India. The ambivalent meaning of the site with its association to colonialism but also to mourning leads to a situation where the site has been ignored so that it has not been destroyed but left 'to slip into a gradual death' (Chadha, 2006). In the chapters of this book, several feelings of ambivalence are present: the thrill of exploration can turn into fear, places of neglect are treated with care and wastelands considered with admiration.

Historian Changbai Wang (2020: 562), who has studied diasporic histories, has used the concept ambivalent heritage in relation to engagements with the physical remains of the past that people in the present are interested in but feel confused about how they should be interpreted and represented. In contrast to dark heritage (e.g. Thomas et al., 2019), rather than being painful, ambivalent heritage confuses people and opens up different interpretations. It produces situations in which parties involved in the heritagization processes become perplexed by complex questions of 'how to interpret the past, by whom, and for what purposes'. The physical qualities of the sites themselves seem to play an active role in their interpretation (Wang, 2020: 578): in the case of postindustrial structures, their magnitude and perceived emptiness invites ambivalent meanings. At postindustrial sites, the abandoned state of some of these spaces fortifies the ambivalence. It must be noted, however, that whereas an unused industrial site may seem abandoned for an outsider, such as an urban explorer, it can be a potential heritage site or a store space for the owner or for the locals. Ambivalence, as a process, disrupts the heritage-wasteland dichotomy and helps us to understand spaces in between.

Ambivalent heritage is, similarly to all cultural heritage, closely linked to collective identities, and can thus have multiple meanings to different communities who have their own interest in history and the re-use of heritage. In defining cultural heritage, we take its process-like nature as a starting point. To recognize something as cultural heritage is a product of a process of valuing and selecting. In this process, the meanings adhered to it are continuously created, recreated and validated for the present situation (Smith, 2006; Macdonald, 2013). Archaeologist Rodney Harrison (2015: 28) has defined heritage as a material-discursive process in which the past and future arise out of dialogue and encounter between multiple embodied subjects in (and with) the present. This aspect underlines that heritage is dynamic, and always political and emotional (see also Tolia-Kelly et al., 2017; Wetherell et al., 2018). Hence ambivalence is one aspect of the process called heritage, tying it closely to aspects of identity, emotions and politics.

Postindustrial spaces are spaces left over from their use in industrial production. In this book, the examples of these spaces vary from old factories to mines, and geographically from Finland, Sweden, Estonia and Wales to Ukraine. As already mentioned, these spaces are often defined by their complexity and huge size, which makes it difficult to protect or re-use them. Most industrial areas are tied to the surrounding landscape by a network of infrastructure, that enables the distribution of products and materials, such as storage, railways, roads and harbours. The infrastructures of industrial communities also includes various forms of services, including social public institutions, such as schools and hospitals, as well as trade and leisure activities. When production in the industrial plant stops, some spaces and infrastructure might be re-used for new purposes, while others are not needed anymore, and their spaces become abandoned. Some of the spaces explored in this book are left to ruination, others are restored to heritage sites and some are taken into new (industrial or other) use. They can also be in the process of turning from one to another or in an unknown liminal stage of waiting for potential new use, such as tourism.

The ambivalence can also be seen in sites that are not directly related to postindustrial sites but to the transforming urban and rural landscapes, such as in the two cases from Finland in this volume: the suburban area of Korkalovaara in Rovaniemi, as well as the Paimio Sanatorium, one of the most internationally renowned Finnish buildings designed by the functionalist architect Alvar Aalto. The specific purpose for which this vast building was built makes re-use difficult. In addition to these urban and rural sites, one text brings to the fore a curiosity that challenges heritage values based on rational scientific (historical and archaeological) knowledge, the so-called Temple of Lemminkäinen, a cave with contemporary folklore about hidden treasures. These sites help us to understand the concept of ambivalence from yet another perspective, as they have, alongside valuation, invited negative, contradictory and doubtful emotions.

In this book, we discuss multiple meanings given to industrial heritage in postindustrial times, and the experiences and emotions connected to material heritage sites. We will scrutinize the present state and future scenarios of different industrial sites and infrastructures as ambivalent heritage, considering the values adhered to industrial heritage and its multiple layers. The authors of the book represent a variety of fields, including heritage studies, archaeology, ethnography, sociology and tourism research, hence giving a multidisciplinary view to the different ways ambivalent heritage is experienced and used today.

Approaches to processes of deindustrialization

Industrialization has affected different countries and even regions within countries more profoundly than others. Resulting from this, deindustrialization has also been more dramatic in regions with heavy industry. Deindustrialization is a complex process with varying degrees at different places. The impacts of deindustrialization on community can also vary at different times (Mah, 2012; Wicke, 2018: 1). The study of regions instead of nations has been seen as a more useful spatial category in the historical studies of industrialization and deindustrialization (Wicke, 2018: 1), but there is a need to also study areas peripheral to industrial heartlands and outside the main fields of industry that were often male dominant, such as coal and steel (High, 2022: 464).

Industrialization and deindustrialization have multiple effects on local communities. These processes do not only affect infrastructure and economy, but also class and gender relations, migration, language and memories (Wicke, 2018: 1). Hence the studies in industrialization are multidisciplinary in nature, including, for example, history, ethnography, heritage studies and archaeology. Deindustrialization studies have their roots in the United States in the late 1970s and early 1980s, where they were tied to industrialization processes (Berger, 2023: 253). The development of postindustrial society and the loss of jobs related to heavy industry gave rise to a need to understand this transition. From the beginning there has been a special emphasis in oral histories and ways the industrial past is and should be remembered, as these memories were tied to local identities (e.g. Berger et al., 2022; Berger, 2023: 253). On the other hand, there has been an emphasis on the materiality of (de-)industrialization with industrial archaeology focusing on the materiality of manufacture in its early stage, but later scrutinizing also how material culture can reveal the changes in the deindustrializing societies, and their social and cultural meanings. Heritage studies again has turned the focus from tangible to intangible industrial heritage, including the knowhow and remembrance of industrial work (e.g. Orange, 2016).

For former workers and other members of the industrial communities, industrial heritage often carries *nostalgic* values. In many cases, factory work offered workers stability, daily rhythm and income, and a social community to belong to. When reminiscing about their daily work at factories, former industrial workers often emphasize the sense of community (e.g. Koskinen-Koivisto, 2014). Nostalgic reminiscence, however, often gets mixed tones when the industrial communities are affected by deindustrialization (e.g. Clarke, 2015). At the local level, deindustrialization causes varying, often strong emotional responses. When industries are stopped and factories close, the feelings of pride over manual work and daily toil change into uncertainty. Diminishing communities and disappearance of familiar environments often cause deep feelings of loss and sadness (e.g. Koskinen-Koivisto, 2014). Deindustrialization has been interpreted as a cultural process of displacement that moves industrial workers to the cultural periphery (Manuel, 2015: 169).

The recent post-humanistic turn has challenged researchers of heritage studies to deconstruct the dominance of anthropocentric thinking and to reflect on more-than-human aspects as well as future-reaching thinking of heritage practices (e.g. Harrison et al., 2020; Sterling, 2020). Industrial heritage provides a fruitful ground to these reflections, as industries reshape and transform the environment, often leaving behind huge industrial constructions, waste, toxics and other infrastructures that continue to exist and influence the local landscapes and ecosystems. Extreme cases of extractive industries and nuclear power may leave the territories poisonous, even lethal. The afterlives of problematic industrial sites bring to the fore challenging questions related to sustainability, conservation and management, calling for a need to address the industrial heritage from the perspectives of transformative pedagogy (e.g. Mezirow, 2018; Viita-aho et al., 2023).

New uses and values

Even when industrial work stops and people move after new job opportunities, the postindustrial landscape is filled with remnants of the industrial past. Apart from abandoned or leftover structures, these can include, for example, narratives, practices and ecological impacts (Jaramillo and Tomann, 2021). Sometimes the remnants of the past can be presented with marks that are not easily recognized for those unaware but still carry pride for the past. In the Renforsin Ranta business park, in Kajaani, Finland, the former United Paper Mills (UPM) factories have gained new use for ICT, technology and energy companies, but the old logo of the paper factory as well as previous structures are still visible because of efforts of a keen individual working there who wanted the past to remain visible in the current use-phase of the structures (Figure 0.1). At the same time the older factory area has proved difficult for new use as it is not suitable for modern industrial use, and its location within the monitored modern industrial area restricts its use for, for example, cultural events.

The Renforsin Ranta is one example of a former industrial site that has been taken into new use. The new users may want to document and save some of the memories from the past industrial glory, as is the case in Kajaani. Some sites may even receive a heritage status, which can positively contribute to the development of the place from which the industry has left (Riesto, 2018: 35). In Chapter 2, Haanpää et al. discuss the possibilities of seeing a suburban area in Rovaniemi, Northern Finland, through the lens of heritage tourism. But the sites can also mean a lot without an official heritage label. For the communities and individuals that exploit and use the outdoor space yard or industrial landscape for their activities, it may not even matter who the current owner of the industrial site is or what their plans for the site are.

Several authors in this volume raise the issue of whether postindustrial sites could be used for tourism, or scrutinize the different ways tourists could experience these sites (Haanpää et al., Chapter 2; Ojala, Chapter 5; Pashkevich and Brandt, Chapter 7). The development of sustainable tourism requires that the ecological, social, cultural and economic dimensions are taken into account as a whole. A key question for cultural tourism in local destinations is whether the different actors share a common vision of the industrial heritage of the destination. Ultimately, it is largely about the values attached to the industrial heritage. For example, gender issues need to be opened up. Views of the industrial heritage as the work of men narrow down the culture and its representation. It is important to discuss in the local settings which signs and symbols of industrial activity are selected to represent the past and industrial identity in the destination. Furthermore, the question of who should determine the choices and their interpretations and meanings is essential. Both tangible and intangible heritage have significant roles by converting heritage sites: whether local experiences, memories and identities are part of a positive industrial heritage that is considered in a pluralistic way. For many reasons, not all postindustrial communities are interested in receiving commercial tourists. As the chapters in this volume prove, in some cases commercial tourism can even hinder other

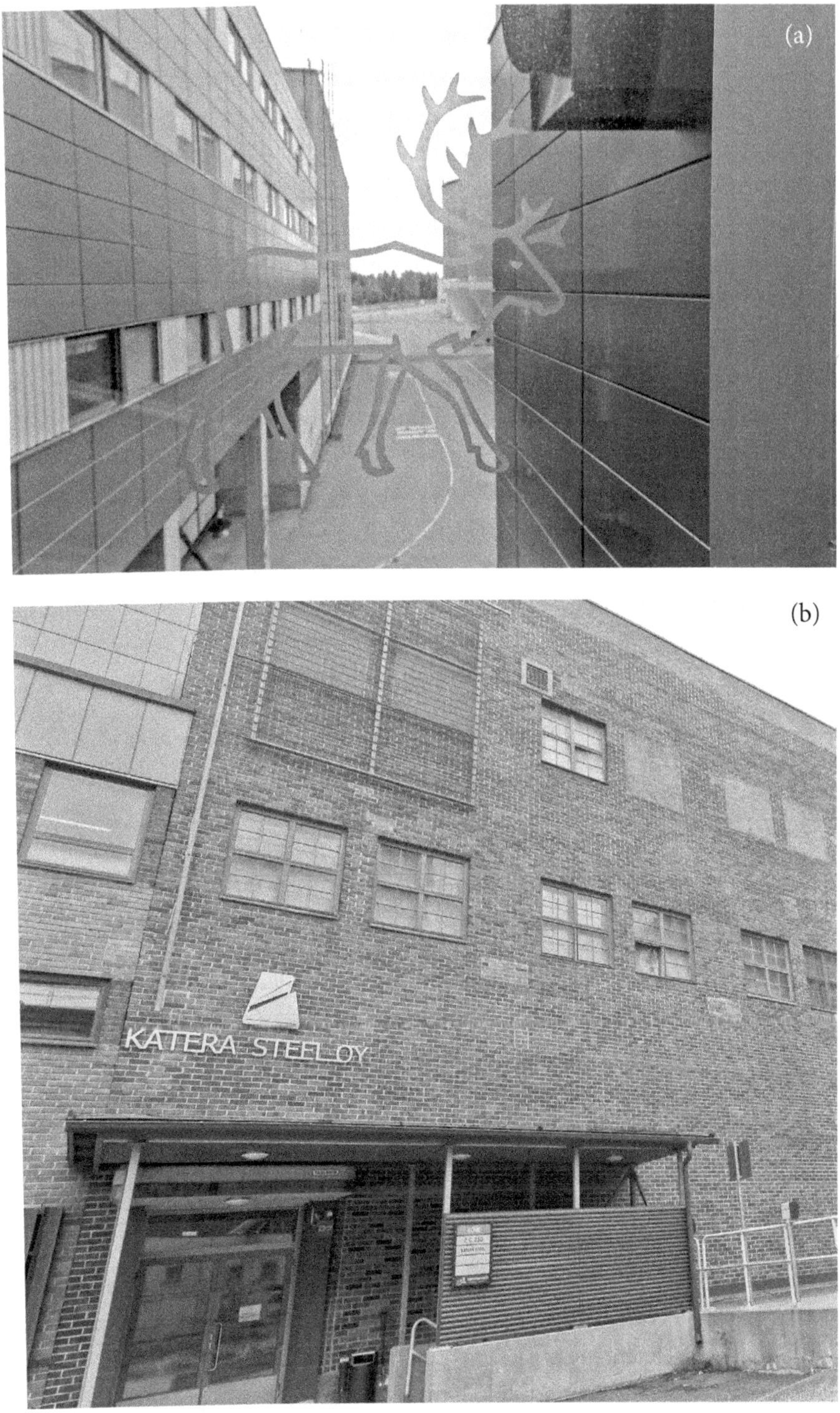

Figure 0.1 (a) The old logo of UPM as a window decoration and (b) Places of old windows marked with different tiles (Photos: Ilja Koivisto, 2021).

ways to use the sites. Volunteers and other informal networks working alongside or initiating formal actors in the meaning-making heritagization processes create the necessary conditions for a multi-voiced approach (e.g. Spiwak, Chapter 1). However, it is necessary to note that some postindustrial sites, because of their location and history, are more difficult to convert into visitor destinations, even if they are of interest to the tourism sector, while others are officially protected (Cercleux et al., 2022: 65; Mildeberg and Vider, 2022: 71; Della Lucia and Pashkevich, 2023).

Many sites in this book do not fall into the category of administrated tourist attractions but they attract individuals on their own search for experiences. As Veera Ojala (Chapter 5) demonstrates with the case of Chernobyl, personal experiences and the feel of freedom can be important for tourists and not found in formal settings. Albina Pashkevich and Daniel Brandt (Chapter 7) use the concept of 'experiencescape' as they show how tourists are not only visiting the mine but the area around it. This brings forth the industrial landscape as a network of infrastructure, waterways and the surrounding nature. The Paimio Sanitorium case (Mäki and Ruotsala, Chapter 3) highlights the importance of the surrounding area – in Paimio, the forest – for the heritage site.

Abandonment and alternative engagements

The processes of deindustrialization often lead to situations where sites become abandoned or are somehow problematic as their use is non-decided and ambivalent, as noted by Martínez (Chapter 9). But the abandonment can also be seen as an alternative approach that may bring to the fore new aspects for the future of these sites. Abandonment can open possibilities for use to different groups of people and individuals. In Chapter 8, Äikäs et al. describe these possibilities from the viewpoint of affordances. The structures, atmosphere and visuality of old industrial sites afford modes of interaction for hobbyists using these sites. This is also evident when old mines are used for outdoor activities (Pashkevich and Brandt, Chapter 7) or for liminal explorations (Martínez, Chapter 9).

Abandonment is a characteristic that is strongly associated with old industrial structures. This is evident, for example, in the hashtags that are used in association with these sites in social media, such as Instagram (e.g. Matila et al., forthcoming). But despite the seemingly abandonment of some sites, they are usually owned properties with owners who might have their own ideas of appropriate use. In the University of Oulu, researchers of the project 'Smokestack Memories' interviewed owners of old factories who saw abandonment as a problematic label for various reasons. It invites people to trespass, hence possibly causing harm both to the building and to themselves if the building is in a state of ruination that can lead to dangerous situations and accidents. This is especially evident as some of the hobbies using old industrial sites are connected to risk taking and adventure (Äikäs et al., Chapter 8). Other owners stressed that the seemingly abandoned factories were actually in active use, for example for storing boats during the winter. There were also some plans to turn old factories into heritage sites as they still had their working machineries and tools intact (Figure 0.2). This leads to complex questions of ownership and responsibility.

Figure 0.2 Machineries at the Kuhmoinen sawmill, Finland (Photo: Ilja Koivisto, 2021).

The abandonment of these sites is further problematized by the fact that they are inhabited by plant and non-human lifeforms. During our fieldwork we met pigeons, seagulls, a fox and traces of moles. The lack of human control at some abandoned places enables plants and animals to take over the space (Edensor et al., 2012: 66). Cultural geographer Caitlin DeSilvey (2017) challenges our thinking of conservation by suggesting collaborating with natural processes. Here, the taking over by nature can be seen as a welcome part of the life cycle of the buildings. The abandonment and the associated ruination also offer lure and emotional attraction (e.g. Edensor, 2005).

On the other hand, plants can be used in the visual imaginaries of the leftover industrial sites to contrast the decaying buildings and growing plants, hence distancing the people and concealing social problems caused by deindustrialization (Storm, 2014: 114). The aestheticization of industrial remains into ruin has been problematized as it hides the problems of deindustrialization in plain sight, turning them to 'fetishized scenes' and to 'ruin porn' (e.g. Storm, 2014: 118; but see also Pétursdóttir and Olsen, 2014). Historian Steven High (2022: 468) has emphasized the need to 'shift our attention from ruins to ruination: a political process that has often left working-class communities impoverished and stigmatized'. 'Ruin porn' can, nevertheless, also comment on contemporary society. Art historian Dora Apel (2015) has connected the ruin imaginary of Detroit to a growing cultural pessimism and a loss of faith in progress, where the aesthetic distancing of representation helps people to cope with their fears. Similarly, in Chapter 4, Tuuli Matila brings forth the layers of meanings that photos of industrial remains can carry, connecting personal loss and deindustralization to ruination.

Human geographer Anna Storm (2014) uses the term 'scars of the postindustrial landscape' to refer to a set of traces left on the landscape by industrial production and the feelings and memories associated with them. It can be a mountainside converted into an open pit; a run-down factory; contaminated land; often an abandoned, overgrown or forgotten place. People can live their daily lives surrounded by industrial ruination and produce complicated relationships with them (Mah, 2012). For example, factory smokestacks in the front yard can be seen as an important link to the region's industrial history and occupational identity or safety hazards. Postindustrial spaces are not merely heritage sites to be preserved or sites to be developed or demolished, but they constantly transform and seek their place between the heritage-wasteland dichotomy.

An idea of preserving by not preserving, by letting the sites live their own life with due stages of ruination, decay and destruction (e.g. Arboleda, 2016;

DeSilvey, 2017), can also be applied to postindustrial sites, but locals and the visitors of these sites can have a need to apply their own preservative methods. In Chapter 1, Alexa Spiwak notes how trespassers have been conserving the sites and fixing up infrastructure. Anna Storm (2014: 107) has observed that often the initiative for the transformations does not come from former workers but from people with no previous connection to the site, often newcomers rather than locals. This was also evident in Varjakka, Finland, where an active heritage work was mainly launched by people who had recently moved to the site, even though it also resonated well with the older residents (Äikäs and Ikonen, 2020).

How to approach creative uses

As noted in several of the chapters in this book, heritagization or even preservation are not the only or are not always the best way to ensure the interaction between the sites and their users. As noted by art historian Svava Riesto (2018: 9), many 'post-industrial landscapes mutate without authorized uses and official plans'. They involve materials in decay, nature penetrating built structures. Those landscapes may never become sites for explicit plans but are visited and inhabited by human subcultures, flora and fauna. These landscapes and sites and their layered histories can offer interesting spaces for visitors who actively engage with their physical qualities and aesthetics, exploring and experiencing them independently or guided by other hobbyists. Experiences of encountering and interacting with these former industrial landscapes and architecture are holistic and multisensory, and are not necessarily conscious but *affective*. Affect can be defined as the embodied state and the initial registering of events in bodies and minds, both embodied and semiotic (Wetherell et al., 2018). The semiotic elements refer to the cultural reading and understanding of the elements related to affective experiences. Also, emotions and senses are culturally coded and valued (ibid). The interesting question is how do we value the decay and alterations of postindustrial spaces? How do they affect us and what kinds of uses do they encourage? Herva et al. (Chapter 6) discuss how the ideas of supernatural and otherness connected to the underground, for example in folklore and popular culture, affect our engagement with the subterranean world, including mines both active and abandoned.

Often, affective experiences and the agentive relationship between the site and the visitor are a source of inspiration for art and community-based art projects. Herva et al. (Chapter 6) have shown how art can reveal and comment on feelings and meanings given to postindustrial landscapes. Art can make this landscape visible in a new way (Storm, 2014: 118; see also Nurmi et al., forthcoming). Although the focus here is on postindustrial spaces, it is important to notice that, for example, the impact of the mining industry on local communities can be seen long before the industry becomes active, in the form of a range of emotions associated with a changing landscape, linked to a sense of place and the values that derive from it (see Storm, 2014). For example, Kaija Kiuru, a visual artist living in Sodankylä in Finnish Lapland, has created a work called *Vanishing Landscape* as part of her 'Protected Aapa Mire' series, which deals with the Viiankiaapa nature reserve and the mining plans for the area (https://www.kaijakiuru.net/suojeltu-aapa--protected-aapa-mire). This kind of environmental 'artivism' uses art as a form of activism in addressing tensions between industry and the local community, but above all it brings the locals' relationship between place, nature and future to the fore. As an international trend, art-related events in relation to mining or other industrial plans near or under the mires can be regarded as a weak but intensifying signal (see Latvala-Harvilahti, 2021: 85–6). Weak signals are signs of how the longer-term future is already a part of the present (Dufva, 2022: 110). Mining activities do not run forever, but active years often cause negative impacts on nature (water, land, flora and fauna). Restoring and rehabilitating or not, landscapes transform. The re-use of post-mining regions adds to their cultural meanings and interpretations, including industrial monuments; the milieu may also act as a museum of mining. In the 'Smokestack Memories' project, art is used as a way of bringing together differing views and voices about industrial heritage (Nurmi et al., forthcoming).

Art historian Venda Louise Pollock has commented on how in the transformation processes of industrial landscapes, communities of artists can take ownership of the site in many ways, and by doing so, make apparent the complexity in the dialogues between people and site (Pollock, 2017). Artivism can be seen as a form of memory activism, which refers to a way to raise contested pasts into awareness for achieving mnemonic or political change. This is often done by working outside state channels and by developing 'creative ways of engaging with the past and making it meaningful in the present' (Gutman and Wüstenberg, 2023: 5–6). The above is central to many of the ways of using postindustrial sites that are described in this book. Whether photographed and presented on Instagram, used for various recreational activities or curated by local communities, these sites become meaningful in a new way, stemming from the individual and communal needs in contemporary society. Hence the meanings of postindustrial sites are not cemented but are intertwined with the needs and worries of contemporary society. Their ruined state can reflect war images from Ukraine and serve the needs for sustainable use of leftover spaces. In the process of re-use, the imaginaries of industrial sites reflect the urgent issues of contemporary society and the scenarios of their future.

References

Äikäs, T. and T. Ikonen (2020), 'Public Archaeology and Archaeologists as a Part of the Heritagization of Northern Industrial Sites', *Fennoscandia Archaeologica,* XXXVII: 197–203.

Apel, D. (2015), *Beautiful Terrible Ruins: Detroit and the Anxiety of Decline*, New Brunswick, NJ: Rutgers University Press. https://doi.org/10.36019/9780813574097.

Arboleda, P. (2016), 'Heritage views through urban exploration: The case of "Abandoned Berlin"', *International Journal of Heritage Studies*, 22 (5): 368–81. https://doi.org/ 10.1080/13527258.2016.1153497.

Berger, S. (2023), 'Deindustrialized spaces', in Y. Gutman and J. Wüstenberg (eds), *The Routledge Handbook of Memory Activism*, 253–8, Oxon: Routledge, Taylor & Francis Group.

Berger, S., S. Musso and C. Wicke (2022), *Deindustrialisation in Twentieth-Century Europe: The Northwest of Italy and the Ruhr Region in Comparison*, Cham: Palgrave Macmillan, Springer Nature.

Cercleux, A.-L., J. Harfst and O.-R. Ilovan (eds) (2022), *Culture, Heritage and Territorial Identities for Urban Development.* MDPI. A reprint of articles from the Special Issue published online in the open access journal *Societies.* https://doi.org/10.3390/books978-3-0365-6129-5.

Chadha, A. (2006), 'Ambivalent Heritage: Between Affect and Ideology in a Colonial Cemetery', *Journal of Material Culture,* 11 (3): 339–63. https://doi.org/10.1177/1359183506068809.

Clarke, J. (2015), 'Closing Time: Deindustrialization and Nostalgia in Contemporary France', *History Workshop Journal,* 79 (1): 107–25. https://doi.org/10.1093/hwj/dbu041.

Della Lucia, M. and A. Pashkevich (2023), 'A sustainable afterlife for post-industrial sites: Balancing conservation, regeneration and heritage tourism', *European Planning Studies,* 31 (3): 641–61. https://doi.org/10.1080/09654313.2022.2154141.

DeSilvey, C. (2017), *Curated Decay: Heritage beyond Saving*, Minneapolis and London: University of Minnesota Press.

Donaldson, E. C. (2018), 'Place, Destabilized: Ambivalent Heritage, Community and Colonialism in the Marquesas Islands', *Oceania,* 88 (1): 69–89.

Dufva, M. (2022), 'Toimintaympäristön analyysi: peste ja sen variaatiot', in H. K. Aalto, K. Heikkilä, P. Keski-Pukkila, M. Mäki and M. Pölllänen (eds), *Tulevaisuudentutkimus tutuksi -perusteita ja menetelmiä*, 105–13,

Tulevaisuudentutkimuksen verkostoakatemian julkaisuja 1/2022, Turun yliopisto: Tulevaisuuden tutkimuskeskus.

Edensor, T. (2005), *Industrial Ruins: Space, Aesthetics and Materiality*, Oxford and New York: Berg.

Edensor, T., B. Evans, J. Holloway, S. Millington and J. Binnie (2012), 'Playing in industrial ruins: Interrogating teleological understandings of play in spaces of material alterity and low surveillance', in A. Jorgensen and R. Keenan (eds), *Urban Wildscapes*, 65–79, Abingdon: Routledge. https://doi.org/10.4324/9780203807545.

Gutman, Y. and J. Wüstenberg (2023), 'Introduction: The Activist Turn in Memory Studies', in Y. Gutman and J. Wüstenberg (eds), *The Routledge Handbook of Memory Activism*, 5–15, Oxon: Routledge, Taylor & Francis Group.

Harrison, R. (2012), *Heritage: Critical Approaches*, London and New York: Routledge.

Harrison, R., DeSilvey, C., Holtorf, C. and MacDonald, S. (2020) 'For ever, for everyone…' In R. Harrison, C. DeSilvey, C. Holtorf, S. Macdonald, N. Bartolini, E. Breithoff, H. Fredheim, A. Lyons, S. May, J. Morgan and S. Penrose (eds.) *Heritage Futures*, 3–19, London: UCL Press.

High, S. (2022), 'Toward the Comparative and Transnational Study of Deindustrialization', in S. Berger, A. Die and C. Wicke (eds), *Deindustrialisation in Twentieth-Century Europe: The Northwest of Italy and the Ruhr Region in Comparison*, 416–73, Cham: Palgrave Macmillan, Springer Nature.

Jaramillo, G. S. and J. Tomann (2021), *Transcending the Nostalgic: Landscapes of Postindustrial Europe Beyond Representation*, New York: Berghahn Books. https://www.kaijakiuru.net/suojeltu-aapa--protected-aapa-mire.

Koskinen-Koivisto, E. (2014), *Her Own Worth: Negotiations of Subjectivity in the Life Narrative of a Female Labourer*, Helsinki: The Finnish Literature Society. https://doi.org/10.21435/sfe.16.

Latvala-Harvilahti, P. (2021), 'Environmental Artivism and Future Agency in the Landscape of Mining', *Ethnologia Scandinavica*, 51: 81–101.

MacDonald, S. (2008), 'Difficult Heritage: Unsettling History', in M.-P. Jungblut and R. Beier-de Haan (eds), *Museums and Universal Heritage: History in the Area of Conflict between Interpretation and Manipulation*, 8–15, Paris: International Committee for Museums and Collections of Archaeology and History.

Mah, A. (2012), *Industrial Ruination, Community and Place: Landscapes and Legacies of Urban Decline*, Toronto: University of Toronto Press.

Manuel, J. T. (2015), *Taconite Dreams: The Struggle to Sustain Mining on Minnesota's Iron Range, 1915–2000*, Minneapolis and London: University of Minnesota Press.

Matila, T., T. Äikäs, O. Seitsonen, E. Koskinen-Koivisto and V.-P. Herva (forthcoming), 'Smokestack nostalgia or hypermodern spirituality? Urban exploration in abandoned factories'.

Mezirow, J. (2018), 'Transformative learning theory', in K. Illeris (ed.), *Contemporary Theories of Learning*, 2nd edn, 114–28, London: Routledge.

Mildeberg, S. and J. Vider (2022), 'Soviet Heritage(scape) in Sillamäe: Documenting the Potential in an Emerging Tourism Destination', *Societies*, 12: 127. https:// doi.org/10.3390/soc12050127.

Nurmi, S., T. Äikäs and E. Koskinen-Koivisto (forthcoming), 'Paikkasidonnainen taide teollisuusperinnön merkitysten kuvastajana ja luojana', in T. Kurki, T. Saresma, J. Pöysä, J. Martikainen, S. Jäntti and S. Pöyhönen (eds), *Taidelähtöinen toisinjulkaiseminen – teoria ja käytäntö*.

Orange, Hilary (2016), 'Introduction', in H. Orange (ed.), *Reanimating Industrial Spaces: Conducting Memory Work in Post-industrial Societies*, London: Routledge, Taylor & Francis Group.

Pétursdóttir, Þ. and B. Olsen (2014), 'Imagining Modern Decay: The Aesthetics of Ruin Photography', *Journal of Contemporary Archaeology*, 1 (1): 7–23.

Pollock, V. L. (2017), 'Remembering the Gorbals: Public Art and Memory in the Post-Industrial Landscape', in G. Heeney (ed.), *The Post-Industrial Landscape as Site for Creative Practice: Material Memory*, 1–16, Newcastle upon Tyne, UK: Cambridge Scholars Publishing.

Riesto, S. (2018), *Biography of an Industrial Landscape: Carlsberg's Urban Spaces Retold*, Amsterdam: Amsterdam University Press. https://doi.org/10.1515/9789048524891.

Smith, L. (2006), *Uses of Heritage*, London: Routledge.

Sterling, C. (2020), 'Critical heritage and the posthumanities: problems and prospects', *International Journal of Heritage Studies*, 26 (11): 1029–46. https://doi.org/10.1080/13527258.2020.1715464.

Storm, A. (2014), *Post-Industrial Landscape Scars*, New York: Palgrave Macmillan.

Thomas, S., V.-P. Herva, E. Koskinen-Koivisto and O. Seitsonen (2019), 'Dark Heritage', in C. Smith (ed.), *Encyclopedia of Global Archaeology*, New York: Springer. https://doi.org/10.1007/978-3-319-51726-1_3197-1.

Tolia-Kelly, D., E. Waterton and S. Watson (2017), 'Introduction: Heritage, Affect and Emotion', in D. Tolia-Kelly, E. Waterton and S. Watson (eds), *Heritage, Affect and Emotion: Politics, Practices and Infrastructures*, 1–11, London: Routledge.

Viita-aho, M., A.-K. Koistinen, H. Kosonen, M. Matikainen, J. Turunen and R. Vargas (2023), 'Epämukava, transformatiivinen ja rohkea museo?', *TaHiTi*, 2: 86–92.

Wang, C. (2020), 'Ambivalent Heritage: The Im/Possibility of Museumifying the Overseas Chinese in South China', *Modern China*, 46 (6): 559–84. https://doi.org/10.1177/0097700419878801.

Wetherell, M., L. Smith and G. Campbell (2018), 'Introduction: Affective Heritage Practices', in L. Smith, M. Wetherell and G. Campbell (eds), *Emotion, Affective Practices, and the Past in the Present: Key Issues in Cultural Heritage*, 1–21, London: Routledge.

Wicke, C. (2018), 'Introduction: Industrial Heritage and Regional Identities', in C. Wicke, S. Berger and J. Golombek (eds), *Industrial Heritage and Regional Identities*, 1–12, London: Routledge.

1

'Renegade Care': Conservation, Co-Curation and Heritage Place-Making in Liminal Spaces

Alexa D. Spiwak

Introduction

An unexpected shape lurks in the darkness up ahead of us as mud squelches under our boots. Several hundred feet underground, I catch only glimpses of it through my fogged glasses as our headlamps track back and forth through the murk. This rusty, four-legged apparition, sitting alone in the middle of an open chamber, has been a new addition since my last visit to Cwmorthin slate quarry.

Rhys and Owain,[1] two mine explorers from North Wales, have joined me on a trip underground to show me a pneumatic rock drill which they and a group of other explorers spent weeks painstakingly moving. Its original resting place was four levels above us in a quarried-out chamber that had begun to collapse. Members of this unofficial group recognized that the drill was in danger of being lost forever and devised a plan of action. Over a span of four weeks, Rhys, Owain and a changing roster of other bodies undertook the arduous task of carefully taking the drill apart piece by piece in the cold damp of the slate mine. The work was slow and careful, using blowtorches to gradually warm and loosen off seized bolts. Once apart, pieces of the drill were transported as an assemblage on a rock cart across the span of the quarry – 'we didn't want people to think we were trying to steal bits off of it' – before the individual components were winched down a manway, transported back across the mine and re-assembled in a safer location.

Although the drill is only one of many projects the group has undertaken to conserve and curate these spaces, it is, as a rule, unusual that such activities are done with the permission or even knowledge of the landowner. On their own initiative they have shored up collapses, dug out backfilled tunnels, oiled machinery to stave off corrosion, installed ropes and ladders and cleared debris from staircases in several mines across North Wales. Owain, laughing, labels their work as 'renegade care': they are not supposed to be there, and they are not exactly what you would call trained heritage professionals. The rag-tag group comes from all walks of life, united in part by their interest in Welsh industrial history and the thrill of trespassing underground. And they are not the only ones: the hidden worlds of North Wales' industrial past are littered with instances of home-grown conservation and curation; these acts of heritage place-making, care and storytelling are undertaken by, and intended for, those who also pass through the threshold from daylight to darkness.

This notion of threshold experiences was formalized by the French ethnographer Arnold van Gennep in his discussion of the ceremonies and rituals that marked a man's passage from one stage of life to the next ([1909] 1960). Van Gennep found that such rites of passage, across multiple cultures, consisted of three overarching stages: preliminal separation, liminal transition and post-liminal aggregation, each with their own defining qualities (ibid.). These stages will serve as the structure for this chapter, guiding the reader through the journey of leaving the surface world behind, descending into the darkness of the mines and ascending back out again. Special attention will be paid to the intermediate liminal stage, which Victor Turner (1967: 93), building upon van Gennep's work, defined as a state of becoming, an 'interstructural situation' that results in 'interstructural human beings'. The liminal stage is inherently an ambivalent one, characterized by ambiguity, paradox, transformation and freedom from social norms and hierarchies. Liminal spaces, often theorized in the context of tourism studies and cultural geography (e.g. Shields, 1991; Pritchard and Morgan, 2006; Huang et al., 2018), also possess these attributes, becoming places that are 'betwixt and between' (Turner, 1967: 110), neither-and, 'out-of-time and out-of-place' (Pritchard and Morgan, 2006: 764). They are spaces set free from the constraints of social order, where one can shed the trappings of everyday life to become a *tabula rasa* (Turner, [1969] 1995: 103). This chapter will employ the concept of liminality in both senses, geographical and sociological, to explore the ways in which transformative liminal experiences are facilitated by physical encounters with the past in liminal spaces.

Rhys, Owain and I, huddled around a digital recorder balanced on the drill, are just such 'interstructural human beings' shivering in a liminal space: Cwmorthin is neither fully open to the public nor forbidden ground, and yet it is somehow both; a place that has been inscribed as world heritage, but left largely without formal conservation; a chaotic space where order has nonetheless been imposed and where decay and curation co-exist; somewhere people unwittingly engage in deeply political practices while attempting to escape from the politics of the overworld, a place that is out-of-sight but never out-of-mind. Down here, in the 'fruitful darkness' of liminality (Turner, 1967: 110), communities of explorers, caretakers and vandals have made these spaces their own, curating small collections of artefacts, undertaking conservation treatments and leaving their own names graffitied next to those of quarrymen. The slate mines of Northwest Wales provoke discussions revolving around communities, access and uses of heritage: who has a right to these spaces, and for what purposes? What becomes of liminal heritage when left without formal care? Do acts of community care, undertaken in the dark of the mines, ever see the light of day? As the result of years of situated ethnographic fieldwork, this chapter is a testament to the transformative potential of liminal spaces and an account of the various uses (and abuses) of North Wales' underground heritage, concentrating on three quarries, Cwmorthin, Rhiwbach and Maenofferen, as case studies.

The pre-liminal: Overworld context

Ffestiniog quarries

The quarries in the vicinity of the town of Blaenau Ffestiniog were born from Ordovician slate veins with relatively low angles to the horizontal and significant amounts of overburden, making mining a much more economical option for stone extraction. Collectively known as the Ffestiniog quarries, the various workings clustered around the town accounted for a third of Welsh slate output in the latter half of the nineteenth century (Richards, 2007: 171). Nestled in the shadow of the Manod and Moelwyn Mountains, the region became known for its blue-grey 'Porthmadog Slates', named after the harbour from which they were shipped (Richards, 2007: 177). Rhiwbach had a relatively early start compared to others, as a small digging in the eighteenth century (Richards, 2007: 200), while sporadic extraction at Cwmorthin and Maenofferen did not begin until the early 1800s (Richards, 2007: 193; Isherwood, 2014: 9). Nevertheless, all three of them hit their stride in the mid-1800s, transforming the small village of Blaenau Ffestiniog into 'The City of Slates' (Lindsay, 1974: 136). Reaching their zenith in the 1870s and 1880s with the help of improved transport links, Cwmorthin, Rhiwbach and Maenoferren suffered a slow decline with sporadic spikes in activity throughout the twentieth century. Only a small handful of quarries in the Ffestiniog area are still in operation today, albeit at a significantly reduced capacity.

As of 2021, significant portions of Blaenau Ffestiniog, Porthmadog, the Ffestiniog Railway and the surrounding relic quarries were inscribed on the UNESCO World Heritage List as Component Part 5 of the Slate Landscape of Northwest Wales. Given that the inscription was limited to Gwynedd County, Rhiwbach quarry in neighbouring Conwy County was excluded from the designation. Both Cwmorthin and Maenofferen, however, were included, as were considerable areas of their underground workings. Despite this great honour, funding for conservation and management is scarce, and the scale of the designated areas is immense. Legislation in the United Kingdom is such that the onus for maintaining listed buildings and properties often falls on struggling villages and equally cash-strapped landowners. Lucky applicants may be successful in procuring support from the Heritage Lottery Fund or other grants for conservation work; Cofio Cwmorthin Remembered, a local historical group manned by volunteers, was able to use a small grant to undertake limited archaeological investigations and reinforce the collapsing walls of surface structures between 2013 and 2015.[2]

These days Cwmorthin is leased and managed by Go Below Underground Adventures Ltd., a company that offers a range of adventure tours with names such as 'Challenge' and 'Hero Xtreme', boasting abseils, free falls and their newly-installed 'goliath' zipline. Although the adit entrance is secured with a hefty metal gate and padlock, vetted individuals are given keys which are occasionally circulated among family and friends. Before this system was implemented, access was granted by calling a phone number posted on the adit gate and obtaining the code for a combination lock. Unauthorized foot traffic has been significantly reduced with the key system, in addition to security cameras outside the entrance.

Further east, beyond Manod Mawr, lies Rhiwbach, a relatively remote quarry with its now-ruined quarry village and auxiliary structures. After shutting down in the 1950s, the underground workings were left untended and quarrymen were put to work in forestry, planting trees on the slopes above the village of Cwm Penmachno. Ownership today is split between a private landowner, who has purview over the upper quarry, and National Resources Wales (NRW), who manage access to a lower drainage adit. Go Below also has an ongoing agreement with NRW to take tour groups into Rhiwbach through a padlocked gate, which was installed in the lower adit in 2003. There is, however, unfettered access to the mine from the upper workings, where the steel grates that previously barred off visitors have now fallen into disrepair.

Maenofferen, located between the two, is owned by the Greaves Welsh Slate Company, a private enterprise that also owns nearby Llechwedd quarry.

Llechwedd has been transformed into one of Blaenau's largest tourist destinations, with similar 'deep mine tours', ziplines and mountain bike courses. Sections of its underground workings have also been repurposed for the aging of artisanal cheddar, while new glamping pods have been constructed overlooking the tips above ground. Maenofferen, by contrast, has been left relatively untended, its extensive surface structures, including a smithy, workshop and sizeable cutting mill, left to ruin. Sporadic efforts have been made to prevent access to the quarry's labyrinthian underground, including the recent backfilling of the main incline, but these efforts have had very little effect. Various groups of people, some belonging to a community commonly labelled as 'urban explorers', still find their way underground.

Betwixt and between 'urbex' and spelunking

The group I have been following, however, cannot really be classified as 'urban explorers',[3] a subcultural movement that sprang up in the early 2000s centred on the practice of trespassing in urban spaces, most typically derelict ruins and authorized-access-only infrastructure. The sites they explore are not urban, for starters, and their methodology more closely resembles spelunking than 'infiltration' – more bright cave suits and abseiling than black hoodies and fence-hopping. Nevertheless, the history of these spaces and the objects within them are an attractor, as is the thrill of visiting spaces few have seen or dare to tread. Underground mine exploration is a breed of its own, with its own online community and code of ethics.

Nevertheless, much of the scholarly discourse surrounding urban exploration is still applicable to mine exploration. Anthropologists and geographers took note of the burgeoning practice of sneaking into ruins, storm drains and subway tunnels in the 2000s and 2010s, and began writing extensively about exploring communities and the spaces they visited. Despite warnings against painting all urban explorers with the same brush (Garrett, 2014), much of the literature boils down to a binary division: two groups, which Pablo Arboleda (2016) called 'performative' and 'communicative' explorers. The former group aligns more with Bradley Garrett's (2014) ethnographic experiences alongside high-profile infiltration crews in London, for whom the joy of exploration came from the act of subverting societal rules and reclaiming restricted spaces, in so doing reaffirming their personal autonomy. Insular and secretive, these performative practitioners prefer to keep locations secret, concentrating their efforts on planning and undertaking their often dangerous and highly illegal explorations.

Their exploits have been likened to hacking (Dodge and Kitchin, 2006; Garrett, 2013), hunting (High and Lewis, 2007) and the mass trespasses of the right to roam movement in England (Solnit, 2006). Groups that Arboleda would label as 'communicative explorers', on the other hand, are described by Garrett (2014: 4) as 'geeky middle-aged enthusiasts getting excited about becoming custodians over derelict spaces', a comment made in reference to Luke Bennett's (2011) study of British bunker explorers. In his interactions with 'bunkerologists', a term he created to describe a very specific subgroup of urban explorers, Bennett (2011: 428) found that his study group was united by their enthusiasm for obtaining and sharing new knowledge and the memorialization of heritage. Significant weight was placed by this group on the importance of phenomenological experience, which he described as 'communing with the spirit of the place' (ibid.). Communicative explorers, according to Arboleda (2016), prefer to share photographs, maps and information about the locations they explore, believing that heritage – even derelict heritage – belongs to everyone, and that everyone should likewise have the opportunity to experience the communion described by Bennett.

Both groups have received criticism, including a scathing excerpt in High and Lewis' (2007: 55) book *Corporate Wasteland*: 'these [are] not the sons and daughters of mill and factory workers.... Despite their stated desire to record places before they vanish, urban explorers are more interested in aesthetics than history.' Many explorers are dismissed as thrill seekers who disregard the law and revel in transgression for transgression's sake; Garrett's work (2011a; 2011b; 2013; 2014) and my own personal experience has shown that this is at least partially true. High and Lewis (2007) were at least partly correct about Owain and his comrades: many are indeed not descended from quarrymen. But nobody in North Wales is ever far removed from the slate industry, especially not in Blaenau, where Rhys lives, encircled as it is by tipping piles and derelict quarries. In reality, the two groups – performative and communicative – are on opposite ends of a spectrum, with most individuals falling ambivalently somewhere in the middle. Owain chortles at Garrett's (2014: 4) description of 'geeky middle-aged enthusiasts', adding that it was closer to the truth than he wanted to admit, albeit not quite the full picture. Some members of the group are not really in it for the thrill of subverting laws or the romantic obsession with ruins, but some are, and sometimes it depends on the place or the day. Others are much more concerned with the historical elements of the quarries, delving deep into the distal corners of the mine to find 'untouched' areas. It has become obvious to me that there are elements of both Arboleda's groups present within the mine-exploring community in the United Kingdom, a group that in my experience has been welcoming and enthusiastic about sharing both space and history with others. At the same time, there is an undeniable element of gatekeeping involved, sometimes literally, and the practice, like typical urban exploration, is heavily male-dominated. Some members lean more towards the performative end of the spectrum, cultivating a strong sense of stewardship, sometimes extending to quasi-ownership, towards certain locations. This feeling of possession among one group or another leads to tensions surrounding access, use and maintenance of these spaces, an extra layer of contestation on top of legal and economic considerations.

Despite disagreements between individuals and groups, everyone in the mine-exploring community is united in their desire to seek out places that are unfamiliar, dark and at times dangerous; the appeal of such liminal spaces rests, at least in part, on the fact that they are so uncanny and unfamiliar. To many, they are a welcomed break from ordered society (Preston-Whyte, 2004), a blank page upon which to tell a story of the individual's choosing. For many groups of explorers, these stories revolve around a carefully curated version of the quarry's past.

The liminal: Underground

Co-curation and heritage place-making

Cwmorthin is one such location where curation is a prescient topic, even more so now that it has transitioned to being a for-profit adventure tourism site. Although Go Below tours feature a good deal of historical information, much to the disappointment of some thrill-seeking reviewers on Trip Advisor, some mine explorers disagree with the claim that Go Below is a means of seeing the mines 'in their raw and unsanitized state, as the miners left them'.[4] There is certainly a good deal of material that the quarrymen *did* leave behind, including a wealth of artefacts and equipment that can be found scattered throughout the complex's hundreds of chambers. Although some artefacts have undoubtedly been squirrelled away into backpacks and carried out of the mine, an interesting phenomenon has developed within these underground spaces: the co-curation of small exhibits along shelves and slabs, displays which are dirtier, more haphazardly assembled than those found in museums and devoid of any interpretive text. Co-curation, defined by Brewis et al. (2021: 4) as 'negotiated identification, selection, preparation and interpretation of materials', is often framed in terms of collaborative partnerships between heritage professionals and the general public. The displays that have organically developed in places such as Cwmorthin, however, appear to be entirely community-led and uninfluenced by 'expert' narratives. The assemblages are instead created and staged by the people who visit these out-of-bounds spaces, comprised of the things they find interesting and relatable.

My first encounter with one such Cwmorthin collection was in 2014, on a pair of shelves beside the *caban* (messhouse) in an area known as Floor G. On the first shelf to the right of the entrance was an assortment of old shovel heads, rusted kettles and sherds of marmalade jars and teacups; quarrymen famously subsisted on bread and sweetened tea for most of the working week (Jones, 1981: 32), and the collection here was a direct reflection of their subsistence patterns. Interestingly, however, this assemblage has not remained static, and instead has grown and changed over the years. An explorer's photograph from 2005 depicts a relatively barren shelf with only the shovel blades propped up against the chamber wall. A subsequent visit in 2018 revealed that the 2014 collection had not only expanded to fill the first shelf, but it had spread to the second. By 2023, both shelves were full and objects had spilled out over slate slabs placed on the floor (Figure 1.1). This collection, in particular, shows signs of active curation: objects are grouped together by type, with 'finer' or more intact examples on the top shelves and smaller, miscellaneous pieces on the floor.

Figure 1.1 The *caban* display as of February 2023, Cwmorthin (Photo: Alexa Spiwak).

These co-curated 'exhibits' pop up in unexpected places throughout Cwmorthin, both on and off the Go Below tour routes, sometimes deliberately staged as a form of heritage storytelling. Tucked out of the way in a small corner we found a slate slab arranged with broken teacups, marmalade jar sherds, bottle glass, pipe stems and decaying articles of clothing. A tallow candle was affixed to the wall with a ball of clay, burnt down to a stub. Beside it, on a small outcrop of rock, hung the tattered remains of a hat. The negative impressions of pipe stems and sherds could still be seen in the dirt, as were the finger-streaks of previous visitors digging through the muck for their treasures. Still caked in mud, the staging of these objects brings the image of quarrymen huddled together on their breaks, smoking pipes by candlelight, to life in the darkness.

This act of 'time travel' (e.g. Holtorf, 2017), of transforming space into place by imbuing it with memory and meaning, is a subtle art, one which Owain, Rhys and their friends have also toyed with. Returning to the drill, Owain gestures to the open chamber with outstretched arms. '*This*', he says, 'is not where we originally set it up.' Rhys and Owain explain that they had originally found a less obvious place to reconstruct the drill, turning it towards a rock face to give the appearance that its operator had left work one day and never returned. To them, the context of its use was important, and it was staged as such: 'That was someone's life working on that thing', said Owain, '[It] brings you back to what it was like. That's why we put it where we put it.' Before it was moved, visitors had to crawl over a partial collapse in a relatively remote area to reach the drill, the intent being to reward adventurous spirits with a moment of discovery and surprise. They suspect that the landowner, or members of their team, moved the drill to place it along the path of the underground tours that now filter through Cwmorthin's passageways. A small, grey plastic tag affixed to one of the legs ('This is more delicate than it looks. Please do not move or adjust the drill.') lends credence to their theory. Propped up on bits of waste rock to prevent the long, heavy bit from falling out, the drill looks slightly stranded and out of place so far away from the chamber face (Figure 1.2). Rhys scoffs at the plastic tag – 'this thing is solid as' – and goes on to compare it to a glass case in a museum. Both make comments about the 'safe and sanitized' underground tours at other locations, with Owain adding that emergency exit signs and electrical lighting rob you of 'get[ting] to feel how the [quarry]men felt, seeing things just by the light on your helmet … the isolation they must have felt'. This importance of the embodied experience resurfaces time and time again in scholarly literature on ruins and urban exploring (e.g. Edensor, 2005; Sørensen, 2007; Bennett, 2011; Garrett, 2014), in discussions of heritage tourism (e.g. Smith, 2014; Holtorf, 2017;

Figure 1.2 The drill in its new resting place, stranded in the middle of a chamber, Cwmorthin (Photo: Alexa Spiwak).

Burlingame, 2020) and during my own interviews: both Rhys and Owain agree that the tactile experience of (carefully) handling objects in situ is far superior to the experience of seeing them behind glass.

All the same, visibility is a recurring theme when it comes to ad-hoc assemblages of artefacts. By and large these small displays are along main thoroughfares or near landmarks throughout the quarry, the intention being that they are seen and experienced by as many people as possible. One salient example can be found on the same level where the drill now resides, in an open area with two boilers, an oven and a rectangular concrete base for compressors, situated at the end of a steep and harrowing descent down a manway. The route has now been made slightly safer through the installation of anchored ropes, allowing the manway and the compressor area to be incorporated into the adventure tours that are operated by Go Below. The collection here is influenced by the fact that many stop to use these concrete slabs as benches for their lunch or to calm their nerves after the descent, the old blending with the new: modern contributions include yellowed tea lights and an ammo can that contains a tattered guest book, some pens and emergency candles. The rest of the collection, however, is split between boots, components thereof, tattered scraps of clothing and miscellaneous iron components. The concentration of personal items, such as hats and hobnail boots, in this particular area seems strange until one looks back up towards the boilers and oven; speaking with a former quarryman in

2018, I was told that breaks were often taken in the ambient warmth of the boilers, and it was an opportune place to dry soggy boots and socks. As with the other small collections throughout the mine, it began with the placement of a singular boot and hat, but visitors have continuously added to the compressor base display as items have been found over the years.

This phenomenon is not unique to Cwmorthin, either. Rhiwbach, known for being so remote that workmen lived in barracks even into the 1950s, is another location that hosts groups of students and tourists, sometimes up to sixteen at a time, through an agreement between Go Below and the two landowners. It is not the easiest to get to: *Rhiw bach* translates to 'Little Hill', a name I can only imagine was a tongue-in-cheek reference to the formidable climb up to the workings. So steep was the road from the town of Cwm Penmachno to Rhiwbach that even modern buses could not make the ascent, and quarrymen were forced to walk (Gwyn, 2015: 208). Despite the hike and the mine's famously soggy condition, the lower levels are littered with instances of co-curated displays.

The presence of such large groups of tourists muddles the picture of who, precisely, is curating these exhibits. Given that Cwmorthin's most notable examples are along the most common through-trip routes, could it be paying tourists that are collecting items and grouping them together? Although they no doubt contribute, Rhiwbach tells a slightly different story: the small collections of artefacts are found in many cases where the tour groups *are not*. Rhodri, another member of the group and a tour guide for Go Below, tells me as we slosh through the flooded adit that the tours have a rather strictly set route, both to maximize the 'adventure factor' with climbing and abseiling installations, but also because of the mine's increasingly dangerous condition. The walls of stone that were left in between chambers to prevent collapses, known as 'pillars', have been shaved down, punched through or taken out entirely in many places, weakening their integrity. Fault lines, inclusions and quartz veins have resulted in both large and small collapses, and even the rockmen were seemingly aware of the poor quality of the stone: many chambers on the lower levels are distinctly 'A' shaped to minimize the risk of deadly rock falls.

Several notable displays are found throughout these unstable lower levels, including near the *caban*, just one chamber past where tour groups turn off the main tunnel to paddle an inflatable raft over a flooded section. The usual suspects are lined up on the wall of the partially collapsed structure: marmalade jar sherds, bottle glass and corroded bits of iron. Interestingly, however, someone has attempted to preserve cigarette packets by backing them with black paper and laminating them between two sheets of plastic – a well-meaning act, but one that is irreversible without completely destroying the contents. Similar displays can be found on shelves and benches built from waste rock throughout the lower floor; as we pick through the artefacts, Rhodri laments that the tour groups never get to see any of it. Elsewhere, stacks of crudely split slates point to experimental attempts at learning the quarryman's trade.

Rhodri is one of the few young people that still live in Cwm Penmachno, a former quarry village trundling head-first towards ghost town status. He is sceptical of the UNESCO designation, believing it came 'far too late, and doesn't do anything', but he is nevertheless passionate and knowledgeable about the area's history. As he dances over the rocks on our way up the incline – lovingly nicknamed 'The Waterfall' because it channels a deluge down from the upper quarry – he tells me that he is grateful he can at least make a living for his small family from the mines, even after they have closed. After ascending two floors he veers off the incline down a passageway, leaving me to scramble after him. He tells me that he used to take his 'good tourists' on this side trip on Floor 4, depending on time and group size, to show them a little something extra: a *jympar* (jumper). Before the introduction of pneumatic rock drills like those found in Cwmorthin, holes were drilled by hand using a long, chisel-tipped iron bar with a swelling partway down its length. A quarryman could spend a whole day using repetitive percussive force to gouge out charge holes for a single slab, depending on the hardness of the rock. The *jympar*, which had been in place during my previous visit in 2018, was now missing, and according to Rhodri it had been gone since 2022. Before it was taken, it was a lone object amidst bare slate, inviting explorers to touch and try; the rock beside where it used to stand is still marred from others' largely unsuccessful attempts, none of the holes surpassing more than 2 centimetres in depth (Figure 1.3). Through this tactile interaction, visitors could gain an appreciation for the backbreaking labour involved in extracting slate: the sensation of using such a tool, the way the percussion travels up your arms and chatters your teeth, is impossible to comprehend by simply beholding it. The 'interactive exhibit' had been in place at Rhiwbach for years, each visitor placing the *jympar* back in its hole for the enjoyment of whoever should come next. Rhodri, declining to give a name, claims that someone had taken it to 'fix it up' more than a year ago. 'Shame it's gone', he said, shaking his head, 'You know, it probably taught thousands of people about the slate industry.' With any luck the *jympar* will be returned, allowing visitors – sanctioned or otherwise – to once again step into the hobnail boots of a quarryman, if only for a brief moment.

Figure 1.3 Visitors' previous attempts at using the now-missing *jympar*, Rhiwbach (Photo: Alexa Spiwak).

Stewardship

A few days after our trip down Cwmorthin, Owain and I endeavoured to escape a record-setting heatwave by once again heading underground. Maenofferen, one of the quarries included in the UNESCO nomination, is not open for public access, but it is easy to break away from the mountain bike courses or walk up from Blaenau Ffestiniog in order to evade the trucks that trundle across the property. Perched upon a plateau high above the town, the massive cutting mill and adjacent buildings have begun to collapse into ruin. Large signs in Welsh and English, which warn visitors of the danger of death, fall for the most part on deaf ears. Preventing access to the underground workings of the quarry is nearly impossible as several chambers have broken through to the surface because of a combination of two techniques: they were 'worked home' and 'untopped', meaning that slate has been taken from horizontal levels, driving one long chamber from the flooded depths to the top of the mine, and then opened up from the surface. Even with the main incline backfilled shut, the slate veins at Maenofferen have a gentle slope, meaning that anyone can find an untopped chamber and go underground with ease. Owain and I do just that, following a switchback path worn through the backfill down four or five levels, past rocks spray painted with yellow arrows and over small heaps of litter and beer cans. He lets out an exasperated sigh before muttering 'bloody hooligans' under his breath.

Arriving finally at a horizontal level after scrambling over a few large slabs, Owain and I veer off to the right down a muddy tunnel, leaving daylight behind us. While entering Cwmorthin and Rhiwbach feels like passing through a portal into some secret world, Maenofferen feels like an actively-inhabited space, well-trodden and chaotically messy. The ground is covered in boot prints, and graffiti is omnipresent. Piles of tangled cables and sheathing, burnt black, litter the passages and chambers as we trudge forward; Owain explains that after the quarry shut down for good in 1999, relatively late by Welsh standards, scrappers came in to harvest what little copper wire they could. Since then, the mine has become a hang-out spot for thrill-seeking youth from all over the United Kingdom, drawn by social media posts and YouTube videos. The foot traffic through Maenofferen increased significantly during the COVID-19 pandemic, culminating in an underground rave, which one Blaenau local estimated to have more than 100 attendees.

Owain seemed to grow increasingly agitated as we walked further through the mine, wiping spray-painted graffiti from the walls as he went. I busied myself seeking out and photographing the small ad-hoc displays I came to study (Figure 1.4), but it became increasingly fascinating to watch my friend move through the space with purpose and authority. He would smash thin slates with initials and curse words graffitied on them, then throw the broken bits out of the way into a small pile by the chamber wall. As we moved through the passageways, he used a gloved hand to wipe away spray-painted arrows and symbols, no doubt placed there as navigational aids for visitors moving through the labyrinth. He told me it was something the group did often, otherwise things got 'messy'; subsequent visits with other members confirmed that this was a unanimous opinion. He would move debris out of passageways as if he was tidying his own living room, muttering disapproving curses in Welsh ('*Iesu Grist!*') under his breath. I helped him in his tidying, thinking all the while about how these actions seemed contradictory: we were clearing the pathway for people that 'shouldn't be down here anyway', according to him, or who were unlikely to use the space in ways that he and fellow group members thought were acceptable. As we worked, he told me about how the group would put rock carts back on their rails to get them up out of the mud, but they were always inevitably pushed off again. Not an easy act of vandalism, either, as they take four people to lift. It seemed to Owain that nothing was safe or sacred, and he was not alone in feeling frustrated and exasperated by

Figure 1.4 A small collection of objects on a stone plinth, Maenofferen (Photo: Alexa Spiwak).

what they perceived to be an endless, faltering battle: 'It's a wonderful place that's just been destroyed.' Many members agreed that social media and the internet has made the situation worse, motivating people to commit increasingly stupid acts in front of a camera for the sake of 'likes' and 'content'. Their frustration is born out of real care and concern for the preservation of their heritage, which UNESCO deems to have 'Outstanding Universal Value'. Be that as it may, there was an undeniable irony in watching people so otherwise concerned with safety wipe navigational signs off the tunnel walls, all in an effort to maintain a certain level of historical authenticity (Figure 1.5).

Communitas/communities of practice

Owain and I struggle to catch our breath after hauling ourselves up and out of Maenofferen's main incline and back into the sweltering late-afternoon heat. As we stand, distant voices and the glass-like tinkling of boots on broken slate grow steadily closer, and two bodies suddenly appear alongside us. They are young men, likely in their twenties, one of them with a headlamp and the other wearing

Figure 1.5 Owain 'cleans up' wayfinding graffiti with his boot, Maenofferen (Photo: Alexa Spiwak).

shorts and sandals. They both politely say hello. We wish them luck as they disappear down the incline, nary a helmet or backpack between them. A third man sheepishly appears behind them, trying to hide a lit joint in his curled hand, and informs us that there should be one more coming after him. As he, too, disappears down the incline, the unmistakable smell of marijuana trails behind him. Owain and I look at each other, sharing a gaze of mutual disbelief.

Shaking his head as we stripped out of our sodden caving clothes beside the car, Owain remarks that there will likely be a 'call-out' for Cave Rescue later in the evening. Several mine explorers lend their expertise to the volunteer rescue group, who assist curious humans (and the occasional animal) that find themselves in distress underground. Despite the stupid choices that are often made in advance of such situations, Cave Rescue still goes. According to one volunteer, 'That's just what you do.' Luckily, that call never came. In that moment I was reminded of our lengthy conversation around the drill, where Rhys expressed concern that it was just this sense of communal responsibility that was slowly disappearing: 'People used to take care of each other in Blaenau. It's not like that anymore.' His general feeling was that once the quarries closed, the townsfolk lost the very thing that bound them together.

Victor Turner ([1969] 1995: 126) spent a great deal of time on this concept of community in his later book *The Ritual Process*, a phenomenon that he called 'communitas'. He posited that the dissolution of societal structures during the liminal stage, which would have previously divided individuals, allowed for the formation of intimately bonded groups. A person in liminality 'becomes a *tabula rasa*, a blank slate, on which is inscribed the knowledge and wisdom of the group' (Turner, [1969] 1995: 103), in essence reshaping their identity. Even when one leaves the liminal stage and returns to the structures of everyday life, the bonds of kinship remain; Turner ([1969] 1995: 129) asserted that these ties benefit not only the group members themselves, but the societies to which they returned. In truth, the same sociological factors that bound quarrymen to each other still exist among those who tread in their wake: mutual engagement in a specific activity, a joint enterprise in which they are all invested, and a shared repertoire of specialized knowledge and skills (Wenger, 1998: 74). Together, trespassers like the group I tagged along with, the 'hooligans' and others in between, form communities of practice, engaging with liminal space in ways that are personally meaningful; through these group experiences, individual members become a cohesive unit, and, just as Turner ([1969] 1995) claimed, identities become renegotiated in the process (Wenger, 1998: 149). What separates hooligan from history enthusiast is the source of meaning, but if both groups look closely, they will find much in common: the draw to out-of-bounds spaces, an affinity for the uncanny, the thrill of experiencing liminality and the social cohesion that results.

The post-liminal: Discussion

Resurfacing

Liminal spaces facilitate transformative liminal experiences through which communities of practice are formed and identities are renegotiated. That much I hope to have made clear by this point, but the question remains: was Turner right when he said that such experiences have the potential to positively impact a wider milieu? Does ferreting around in the dark do any good, once everyone has left the mines, dried their caving suits and gone home?

After years of knowing and befriending mine explorers, my biased answer is an emphatic 'yes'. When one member of the group experienced a health crisis, the others rallied around him; when another endeavoured to start an independent artistic project, the rest of the group helped get it off the ground. They support each other through weddings and divorces, births and bereavements. But even outside of this one small group, it seems to me that there is a glimmer of communal care in the quarries, evidenced by things such as local historical societies raising funds for archaeological digs, or instances such as Rhiwbach's laminated cigarette packets. The Maenofferen mill has been the recipient of these small acts of care, with simple scaffolding propping up splintered and sagging crossbeams. A chance encounter at the mill in February 2023 revealed that former quarrymen still kept an eye on the property, patching open sections of the slate roof with hardware store tarpaulins. Even the little co-curated displays underground, never exposed to the light of day, are made with the intention of sharing with others, whether that be by creating little vignettes of the past, unintentionally providing an interactive experience or simply setting up small moments of wonder and surprise. All of it is intended to be shared and managed by other like-minded people, members-in-making of a community, and in this light, it is understandable why the 'hooligans' can be reviled as threatening interlopers. But membership in this seemingly closed community of practice is unwittingly extended to them, too (instances of erasing their spray-painted arrows aside), because they, knowingly or not, experience the small exhibits scattered throughout the quarry. They benefit from the addition of new ropes, anchors and ladders, intended for use by the acceptable form of explorer. And in some small way, they must care, too, because they by and large leave these things alone.

The activities in which these underground conservators and curators participate are not, in and of themselves, revolutionary. They are, after all, still engaging with heritage in a very traditional sense: trying to clear away the clutter, prevent the incursion of the world, stage an 'authentic' version of the past and freeze it in time. Heritage is often produced by and for a cultural elite (Smith, 2006), and this is perhaps where the endeavours of mine explorers – and indeed urban explorers writ large – deviate from authorized heritage discourses. Rather than the typical top-down model of heritage projects being done *to* and *for* communities (Waterton and Smith, 2010: 7), trespassers have rolled up their sleeves and adopted a 'do-it-yourself' attitude. Curating small displays, conserving artefacts or fixing up infrastructure is done *by* and *for* other members of their partially anonymous and widely distributed community of practice. Through engaging with history in liminal spaces, this community is '(re)constructed through ongoing experiences, engagements and relations' (Waterton and Smith, 2010: 8). Community may not be the panacea for all that ails Northwest Wales, but all signs point to positive outcomes from repeated brushes with the liminal.

Uncertain futures

As the Slate Landscape of Northwest Wales garners further well-deserved attention because of its UNESCO nomination, the tensions surrounding authenticity, access and use will have to be addressed. This is a tightrope balancing act for Northwest Wales, as the hands-off approach to management that has resulted in its 'unusually high level of authenticity' and undeniable aesthetic appeal has also allowed the ingress of trespassers and vandals (Management Plan: 57). This issue of access is fraught with nuances. Some trespassers, such as our beloved group and the other nameless curators of the ad-hoc displays dotted throughout quarries, endeavour to have a positive, or at the very least neutral, impact on the sites. Their repeated presence and the arrival of new, like-minded visitors is what creates and maintains the system of co-curation and volunteer conservation that, through the unassuming act of arranging some shoes and teacups on a concrete slab or tidying up passageways, helps to transform seemingly empty space. These actions benefit not just the sites themselves, but everyone who visits them – even those who may not be interested in the history and are simply looking for a thrill. Sealing up the quarries for public safety and preservation, and thereby denying tactile interactions with the past to everyone, is something several locals in the town of Blaenau Ffestiniog told me they did not want to happen. If the marmalade jars and rusty kettles were removed from quarries for the benefit of those who cannot physically visit them, they would likely be perceived as nothing more than junk. All these points harken back to comments made during our conversation in Cwmorthin, shivering while huddled around the drill: *the context matters*. Being in these liminal spaces is necessary for coming face-to-face with the realities of resource extraction, something society has become increasingly separated from.

In a post-Brexit, 'post'-pandemic, high-inflation Wales, where shelves on grocery stores stand bare and empty storefronts line the high streets, there is neither funding nor manpower to do much of anything. More local history in schools, suggested Rhys, but even that requires budgets and curriculum development hours that just do not exist right now. Even if the money could be found, there is still no straightforward answer. Restoring buildings to their former glory would rob them of the romantic ruin aesthetic that draws so many visitors each year. Not every quarry can be turned into an underground adventure park or run controlled, sanitized tours; the county's infrastructure has a carrying capacity for tourism that, in some places, has long been exceeded. Prohibiting access is both financially and practically impossible in most places, and even if it were not, closing people out of heritage sites runs counter to the wishes of Welsh locals and UNESCO's ethos. Management of industrial heritage sites of this type and scale is relatively uncharted territory for UNESCO, and the Slate Landscape of Northwest Wales must therefore forge its own path.

Conclusion

Ambivalence follows naturally from liminality, a state in which many contradictory emotions exist concurrently. This is something I have witnessed repeatedly in my time with these mine explorers: their paradoxical desire to protect heritage and, simultaneously, proudly share it with the masses; the never-ending conflict between the dislike of tourists and the need for revenue streams; and their desire to see the quarries experienced and enjoyed, but only by certain people for certain uses. The tone of the locals in towns such as Blaenau Ffestiniog is equally ambivalent, with optimism and pessimism expressed in the same breath. Many mourn the loss of the slate industry but have no other choice than to hope that heritage tourism will lead to a brighter future. Even with its UNESCO nomination, the future of the Slate Landscape of Northwest Wales is uncertain.

In the face of such uncertainty, however, I find it encouraging that in the liminal, where people are free to do and be as they please, some choose to be caretakers and storytellers. More encouraging still are the signs that these small acts of care, for heritage and for each other, are finding their way back to the surface. The emphasis on the importance of embodied experience is reflected in the care and curation of these places, and the memories they consequently conjure forth of men labouring in the dark. This network of explorers, my friends among them, facilitate physical interactions with a way of life that is slowly being forgotten as the last generations of Ffestiniog quarrymen die. Perhaps, mused Rhys, if more people spent time underground and saw the traces of men in such inhospitable environments, they may begin to care-enough-about to begin to care-for. Given the transformative potential of such liminal spaces, he might just be right.

Notes

1 All names have been changed to protect the identities of participants/interlocutors.

2 'Appeal to save quarrymens' Cwmorthin terrace, Blaenau Ffestiniog', *BBC News*, 31 October 2013. https://www.bbc.com/news/uk-wales-north-west-wales-24749734.

3 For full disclosure, I have been involved with the urban exploring community since 2005. I explored my first abandoned mine in Northern Ontario, Canada, in 2007. My involvement in the 'scene' has no doubt afforded me a greater level of access to both locations and interlocutors, making the methodology of this chapter largely autoethnographic.

4 'Go Below Challenge – oversold. Tripadvisor review, 1 October 2021'. https://www.tripadvisor.com/ShowUserReviews-g186447-d2097486-r812578615-Go_Below_Underground_Adventures-Betws_y_Coed_Snowdonia_National_Park_North_Wales.html.

References

Arboleda, P. (2016), 'Heritage views through urban exploration: the case of "Abandoned Berlin"', *International Journal of Heritage Studies*, 22 (5): 368–81.

Bennett, L. (2011), 'Bunkerology – A Case Study in the Theory and Practice of Urban Exploration', *Environment and Planning D: Society and Space*, 29: 421–34.

Brewis, G., A. Ellis Paine, I. Hardill, R. Lindsey and R. Macmillan (2021), 'Co-curation: Archival interventions and voluntary sector records', *Area* (London 1969).

Burlingame, K. (2020), 'Dead Landscapes – and how to make them live', PhD dissertation, Lund University.

Dodge, M. and R. Kitchin (2006), 'Exposing the secret city: Urban exploration as space hacking', PowerPoint presentation to University of Manchester Geography Department Seminar, 22 February.

Edensor, T. (2005), 'The Ghosts of Industrial Ruins: Ordering and Disordering Memory in Excessive Space', *Environment and Planning D: Society and Space*, 23 (6): 829–49.

Garrett, B. L. (2011a), 'Assaying History: Creating Temporal Junctions through Urban Exploration', *Environment and Planning D: Society and Space*, 29 (6): 1048–67.

Garrett, B. L. (2011b), 'Cracking the Paris Carrières: Corporal Terror and Illicit Encounter Under the City of Light', *Acme: An International E-Journal for Critical Geographies*, 10 (2): 269–77.

Garrett, B. L. (2013), *Explore Everything: Place-Hacking the City*, London: Verso.

Garrett, B. L. (2014), 'Undertaking recreational trespass: urban exploration and infiltration', *Transactions of the Institute of British Geographers*, 39 (1): 1–13.

Gwyn, D. (2015), *Welsh Slate: The Archaeology and History of an Industry*, Aberystwyth: Royal Commission on the Ancient and Historical Monuments of Wales.

High, S. and D. W. Lewis (2007), *Corporate Wasteland: The Landscape and Memory of Deindustrialization*, Ithaca: Cornell University Press.

Holtorf, C. (2017), 'Introduction: The Meaning of Time Travel', in B. Petersson and C. Holtorf (eds), *The Archaeology of Time Travel: Experiencing the Past in the 21st Century*, 1–22, Oxford: Archaeopress.

Huang, W.-J., H. Xiao and S. Wang (2018), 'Airports as liminal space', *Annals of Tourism Research*, 70 (1): 1–13.

Isherwood, G. (2014), *Cwmorthin Slate Quarry*, Merioneth: Peter Burgess.

Jones, R. M. (1981), *The North Wales Quarrymen 1874–1922*, Cardiff: University of Wales Press.

Lindsay, J. (1974), *A History of the North Wales Slate Industry*, Newton Abbot: David and Charles

Preston-Whyte, R. (2004), 'The Beach as a Liminal Space', in A. Lew, C. M. Hall and A. Williams (eds), *A Companion to Tourism*, 249–59, Oxford: Blackwell.

Pritchard, A. and N. Morgan (2006), 'Hotel Babylon? Exploring hotels as liminal sites of transition and transgression', *Tourism Management*, 27 (5): 762–72.

Richards, A. J. (2007), *A Gazeteer of the Welsh Slate Industry*, Pwllheli: Llygad Gwalch.

Shields, R. (1991), *Places on the Margin: Alternative Geographies of Modernity*, London: Routledge.

Smith, L. (2006), *Uses of Heritage*, London: Routledge.

Smith, L. (2014), 'Visitor Emotion, Affect and Registers of Engagement at Museums and Heritage Sites', *Conservation Science in Cultural Heritage*, 14 (2): 125–32.

Solnit, R. (2006), *Wanderlust: A History of Walking*, London: Verso.

Sørensen, T. F. (2007), 'Special Reviews Section: Urban exploration as archaeological engagement: a review of http://infiltration.org/ – "The zine about places you're not supposed to go"', *European Journal of Archaeology*, 10 (1): 89–91.

The Slate Landscape of Northwest Wales [SLNW] (2021), 'Property Management Plan 2020–2030'. Available online: https://whc.unesco.org/document/184595.

Turner, V. (1967), *The Forest of Symbols: Aspects of Ndembu Ritual*, Ithaca: Cornell University Press.

Turner, V. ([1969] 1995), *The Ritual Process: Structure and Anti-Structure*, Hawthorne: Aldine De Gruyter.

van Gennep, A. ([1909] 1960), *The Rites of Passage*, Chicago: University of Chicago Press.

Waterton, E. and L. Smith (2010), 'The recognition and misrecognition of community heritage', *International Journal of Heritage Studies*, 16 (1–2): 4–15.

Wenger, E. (1998), *Communities of Practice: Learning, Meaning, and Identity*, Cambridge: Cambridge University Press.

2

Suburban Heritage as a Tourist Attraction?

Minni Haanpää, Maria Hakkarainen and
Anu Harju-Myllyaho

Introduction

In this chapter we contemplate ambivalent heritage in relation to tourism, and more specifically in relation to an established tourism destination's offering. Our empirical setting, the city of Rovaniemi, is a popular tourism destination in Finnish Lapland. It is well-known because of its main marketed attraction, Santa Claus, and the tourism offering of the city concentrates on different activities around snow and Christmas. Tourists engage in activities such as hunting for the Northern Lights, snowmobile safaris, visiting Santa Claus, and riding reindeer sleighs and dog sleds. In 2022, there were 413,000 foreign tourist overnights in Rovaniemi. In December 2022, Rovaniemi hosted international tourists from forty-five different countries counted by registered overnights. The biggest market areas were the United Kingdom, France, Italy and Spain (Visitory.io).

The current form of tourism in Rovaniemi has its roots in the ideas of journalist Niilo Tarvajärvi, who envisioned a Land of Christmas to be built in Lapland after visiting Disneyland in the United States in the 1950s. The idea did not take off for a few decades, but it remained in different discussions and, in the 1980s, the time was finally right: various national and international stakeholders decided that Santa Claus would become a character to promote Finnish tourism. The national flight company Finnair forcefully marketed Lapland in Britain, and at the same time the Santa Claus Village started to be built in the Arctic Circle, some 10 kilometres away from the Rovaniemi city centre. This resulted in the launch of charter Concorde flight routes to Rovaniemi from Britain in 1984 (see Hakkarainen and Satokangas, 2022: 21; Ilola et al., 2014: 11–15). Christmas and winter tourism began to grow steadily from there onwards and, increasingly, foreign tourists wanted to visit Santa Claus in the Arctic Circle annually.

As tourism has returned to the path of growth after the Covid-19 pandemic in Finland and internationally, the demand and meaning of authentic tourism offerings has also been on the rise. Although Santa Claus is the unquestioned main attraction in Rovaniemi, authentic tourism offerings – specifically the 'live like a local' trend – have gained ground among contemporary tourists (Euromonitor, 2014; Palauskaite et al., 2017). The international trend in tourism is also to experience the 'rougher' or more ragged local phenomena and attractions. Tourists want to gain experiences about local, everyday life, not just attractions designed for touristic use. On the other hand, sustainability and responsibility as concepts and practice are ever more familiar to many tourists, and this results in them being aware and caring about the impacts of tourism on the life of locals (e.g. Dredge and Hales, 2012).

Observing the development of this international trend towards seeking local authenticity in tourism, and with all of us living and having acted as resident activists in the suburban neighbourhood towards the end of the 2010s, discussed in this chapter, we began to wonder whether a local suburb could be a space for tourism. What would this mean inside local tourism and what could it bring and to whom? We set up a project that aimed to explore the northern suburb as a home and in a wider context, as heritage and culture in Lapland. We also wanted to analyse these in relation to Lappish cultural heritage and the image of Lapland as seen through tourism. The background idea of the whole project was based on the ambivalence of suburban heritage in Lapland (and in general): suburbs are not commonly recognized as heritage even though they have a history of more than fifty years, and in Finland around 1.5 million of the population of 5.5 million people live in the suburbs (Ympäristöministeriö). This makes them, besides a remarkable contemporary phenomenon, also a large part of Finnish cultural heritage.

This chapter is structured as follows. First, we discuss the relationship of cultural heritage and tourism in Finnish Lapland and, more specifically, in Rovaniemi. Then, we move on to deliberate the history and present-day suburbs in Finland and how they can be considered as cultural heritage. From there, we move on to consider what tourism and the suburb could have in common, and if and how the suburb can be a space for tourism. We do this by taking part in the discussions on social sustainability and inclusive tourism inside urban and tourism studies (e.g. Dredge and Hales, 2012; Scheyvens and Biddulph, 2018; Wallin, 2023) in relation to our empirical context: the suburban neighbourhood of Korkalovaara in Rovaniemi, Finland. We conclude the chapter by discussing how different values behind tourism development produce and affect what is considered 'valuable' in tourism and what ambivalent heritage as a tourist attraction could mean. This discussion on the one hand can destabilize the way suburbs are understood as cultural heritage (or not) but, on the other hand, the suburban cultural heritage has the potential to bring relevant perspectives to the traditional tourism discourse in Lapland. As Chadha (2006: 341) describes, there is discomfort in ambivalent heritage, which occurs in the dichotomy between the past and the present and in the conflict between different meanings.

The ambivalent relationship of tourism and heritage in Lapland

In essence, heritage tourism is about utilizing the past in the present to promote places and regions to tourists (Berkenbosch et al., 2022: 328). We define heritage here according to Ashworth and Graham (2005, in Berkenbosch et al., 2022: 328), who state that it is 'part of the past, which is selected in the present for contemporary purposes'. However, when dealing with ambivalent heritage, this selection is contested. The contemporary representations are continually altered for different reasons and by different groups. The narratives and heritage that are valued and understood as important are culturally embedded (see Berkenbosch et al., 2022: 328). According to these ideas, the role of heritage in Rovaniemi tourism can be seen as ambivalent.

When considering Lapland and Lappish heritage, the first image might not be a city or a town, let alone a suburb. The imagery is, rather, filled with nature and its phenomena. In the city of Rovaniemi, winter and Christmas tourism imageries form a particularly strong norm (Herva et al., 2020; Ilola et al., 2014). Being in nature and in the wilderness, acting – surviving – form a powerful thread in the narrative of the Lappish heritage. Ethnologists, anthropologists, wilderness wanderers, hunters and fishermen have collected and produced both the heritage knowledge of northern Ostrobothnia and of the Saami people (e.g. Järvinen, 1932; Paulaharju, 1963). Again, the existing tourism imagery in Lapland has taken roots and settled during the past decades to lean on images, discourses and structures around Christmas, winter, wilderness and mystics (see Table 2.1). The stories of the people of Lapland focus mainly on the activities in nature and villages.

Finland's EU membership, which was acquired in 1995, has since contributed to the integration of village history and heritage in tourism development in Lapland. After joining the EU, the regional development funds (e.g. ERDF, Leader and ESF) directed to Finland have financed a wide range of development

Table 2.1 The timeline of tourism development, and the representation of local heritage in relation to it, in Rovaniemi

1900s	Ounasvaara Fell a popular place for celebrations. Domestic tourism to see the Midnight Sun.
1920s	Rovaniemi as a gate to Lapland at the crossroads of the main roads and railroads in the region.
	Arctic Circle signpost becomes a place of stopping and photographing for people travelling the route towards the Arctic Ocean.
1927	First national winter sports competitions in Ounasvaara Fell.
1936	Hotel Pohjanhovi, the first hotel in Rovaniemi, is opened. It becomes a renowned meeting place for tourists and locals.
1944	Rovaniemi is almost completely destroyed in the Second World War by retreating German soldiers. Over 90 per cent of the town's buildings are burned to the ground.
1945	The reconstruction after the war begins. The plan for the town is designed by architect Alvar Aalto and it resembles the shape of a reindeer head.
1950	Eleanor Roosevelt visits Rovaniemi as a United Nation's Ambassador of Goodwill. Roosevelt Cottage is built in the Arctic Circle to honour her visit.
	After the visit, the cottage acts as a summer café selling souvenirs and offering Arctic Circle special postal stamps and postcards. Place of visit for ordinary tourists as well as notable persons and heads of state.
1951	Local Heritage Association Totto is founded. Its missions are upkeeping the home region spirit and enhancing the awarness of the home region.
The beginning of 1960s	A Lappish family, reindeer and Lapp traditional buildings are established in the Arctic Circle for tourism purposes.
	The amount of tourists visiting Rovaniemi increases steadily. They are mainly stop-over visitors travelling by car in Lapland or to the Arctic Ocean and Norway during the summer months.
1967	Based on the ideas of the journalist Niilo Tarvajärvi, Land of Christmas Ltd. is established. The purpose of this was to advance and build Christmas-themed attractions following the example of Disneyland.
The beginning of 1980s	First British charter tourists arrive to Rovaniemi due to collaboration between the national flight company Finnair and commercial British radio stations.
	First reindeer safaris for tourists organized by a local reindeer farmer. In these programmes the participants would aquire 'a reindeer driving licence'.
	Snowmobile safaris begin in Rovaniemi. Their popularity rose towards the end of the decade and in the 1990s.
1984	First Concorde flight from Britain to Rovaniemi.
1985	Santa Claus Village and Office are opened in the Arctic Circle.

	The city of Rovaniemi begins to use the Christmas destination image in its marketing.
1988	Opening of Santapark, the amusement park of Santa Claus. The designers were British and the end result was criticized for being 'plastic' and Disney-like by tourists and locals. Later, the park was altered towards more local Christmas traditions and activities.
The end of 1980s	Christmas tourism becomes its own tourism season. The tourists mainly consist of the British. Besides Santa Claus the marketing imagery of Rovaniemi is based on snowy landscape scenes of Lapland.
1992	The Arktikum Science Centre was opened to the public. The centre also houses the Regional Museum of Lapland. It is one of Rovaniemi's most popular attractions.
1995	Finland joins the European Union. The era of different development projects funded by the EU begins. Several tourism development projects and measures take place and continue until the present day.
2000s	The development of Christmas and winter tourism continues.
2010	Rovaniemi aquires EU Trademark 'The Official Home Town of Santa Claus'. Christmas as a tourism season strenghtens continually. Majority of tourists are international.
The beginning of 2020s	Local Heritage Association Totto becomes active in developing heritage tourism in its Rovaniemi-based museum and the villages surrounding the city.
2023	During the Christmas and winter seasons, there are over thirty direct flight routes and hundreds of charter flights to Rovaniemi from over forty-five countries worldwide. Main tourist attractions are hunting for Northern Lights, snowmobile safaris, visiting Santa Claus and riding reindeer sleighs and dog sleds.

(*Sources*: Hakkarainen and Satokangas, 2022; Ilola et al., 2014; Komppula et al., 2009; Totto).

projects, including tourism development. Often the tool has been the collection of local stories and histories (e.g. Hakkarainen and Leskinen, 2001). For the locals, these have been meaningful activities that have increased their sense of identity. From the tourism perspective, however, the story collections have been more an act of gathering cultural history than tourism development and there is still a lot of work required to turn these stories into (commercialized) tourism products. From the tourism industry's perspective, the identification, recording and collection of local history and heritage can therefore be seen as the first step in the development of tourism. In recent years, the role of the heritage of northern Ostrobothnia and the villages surrounding Rovaniemi city has gained more attention in the development of tourism. There are various ongoing village tourism projects and the Local Heritage Museum is actively developing its exhibition and activities.

Even though there have been a few urban-based stories from Lapland in popular culture outlets (for example the novels written by Finnish author Jari Tervo and the movie *Aurora*, which was released in 2019 and is situated in Rovaniemi city), for the most part these narratives of urban locality and city as a place of living have been left in the margins in Lappish stories. On that account, the ambivalence discussed here is located on one hand in the understanding of suburban cultural heritage (or lack thereof), but on the other hand also to tourism inclusion and the representations of tourism. There has been previous research concerning ambivalent heritage, but fairly little in terms of urban development and tourism. For instance, a plethora of research has been made in the field of indigenous cultures and colonialism (e.g. Chadha, 2006; Donaldson, 2018). However, ambivalent heritage can be found in places such as an undesirable neighbourhood, where the trauma might not be as explicit as in the scenes of colonial history or holocaust, for instance.

Donaldson (2018: 69) asks if affirming and supportive place-making can occur in uncomfortable or painful places. According to her, in the context of indigenous cultures, the basic idea is that heritage places are a solid source of self-determination and strength, but many of the places have been the stage for trauma and loss and, thus, the places can never be the same. Even so, they can provide a means for shared connection and culture. Chadha (2006: 348) notes that there are two relative terms in heritage research that have rounded the topic of heritage arising from historical conflict, contradiction or deep contestation. The terms are 'dissonant heritage' and 'negative heritage' (Chadha, 2006: 348). In the study of tourism, such phenomena, which attach to the above-mentioned dimensions of places and culture, are often tied to the concept of dark tourism (e.g. Martini and Buda, 2020).

In Lapland, the discussion on ambivalent cultural heritage is most often and predominantly reinforced by themes related to Saami heritage, the conflicting experiences related to military history and the extensive use of natural resources beginning from the reconstruction era after the Second World War. In Rovaniemi, local history related to the Lapland war can be seen as a reflection of ambivalent heritage and also of dark tourism (Koskinen-Koivisto, 2023). The inappropriate use of the Saami culture in tourism in Lapland remains a topical issue and creates friction between the Saami people and the tourism industry (e.g. Saari et al., 2020). The devastation caused by the war in Lapland and the loss of entire towns, villages and livelihoods as a result of the construction of hydroelectric power stations resulted in multigenerational trauma among the people living in Lapland. These topics have mostly gone silent and only recently has public discussion and debate on them been possible.

The use of heritage in Lapland tourism has largely been fitted to fulfil the commercial purposes of the industry. Uncomfortable stories, such as the war history, have been mainly left out and, for example, the indigenous Saami heritage has in many cases been used as an element of mystification of the area – in the worst cases without any relation to the actual Saami heritage (e.g. Saari et al., 2020). More recently, the inclusion of northern Ostrobothnian culture as part of Lapland's tourism catalogue has already extended the use of heritage in tourism beyond the indigenous (Saami) and wilderness culture of Lapland. On the other hand, collections of local stories and histories produced in recent decades have contributed to the canonization of the experiences of Finnish agrarian society as normative Lappish heritage. The absence of representations of urban and contemporary everyday life is notable both in the representation of tradition and in the images of tourism. Recognizing and making local urban history visible would diversify the image of Lapland and highlight the elements that have been invisible and silent.

When looking at suburban neighbourhoods in light of tourism, a different type of ambivalence-generating dimension emerges from the perspectives of traditionally presented imageries of heritage in Lapland tourism. Suburban heritage challenges both the norm of heritage-used tourism in Lapland and the prevailing imagery of tourism in the Rovaniemi region. But the popularity and attention towards such marginally located and economically small-scale tourism environments and products that differ from the existing norm has been on the rise during recent years (Paiva, 2023; Palauskaite et al., 2017). Peripheral, urban places have gained the interest of tourists globally. However, research made on the phenomenon is still scarce. From the local point of view, understanding one's own everyday life as culture or cultural heritage is not an unambiguous matter and, as stated before, leveraging such phenomena to the field of tourism poses many ambivalences and even paradoxes. There are a manifold of ethical perspectives to be considered in such an enterprise. We will consider these later in this chapter, but first we will discuss the history of the suburban neighbourhoods in Finland and these areas as part of Finnish industrial and postindustrial heritage.

Development of suburban neighbourhoods in Finland

The development of Finnish suburban neighbourhoods began after the Second World War during societal ruptures. Finland turned from being an agricultural society to an industrial one and, consequently, in many regions this led to a lack of housing. This resulted in the building of suburbs: the apartment building dominated construction that began in the 1960s and 1970s and was typically located in rocky forests on the outskirts of cities. This form of urban planning and housing followed the European development currents, but in a Finnish way, with the strong idea of 'a forest suburb' (see Wallin, 2023: 1–3). The design ideology of the Finnish suburban neighbourhoods was initially based on the 'forest city ideal', in which the buildings were placed freely according to the shapes of the terrain. This was also the case in Rovaniemi, when Korkalovaara was built in the middle of the forest, the construction beginning in the late 1950s.

In general, a move to the suburbs meant increased standards of living, since many people had lived in older houses with poorer living conditions beforehand. Nevertheless, moving to urban places meant radical changes in the way of living for many (e.g. Kortteinen, 1982; Wallin and Rosendahl, 2017: 71–6; Aalto and Laaksonen, 2021: 13–20). As Rovaniemi was growing after the war, the town officials decided to start developing the outskirts of the town and build modern suburban constructs in the area, much as had been done in other Finnish towns and cities. The planning was led by a renowned Finnish functionalist architect, Alvar Aalto, who used the 'garden city' of Tapiola in Espoo, Finland, as a model for the area. Aalto developed an area plan consisting of three different residential entireties, but only one of the original three was realized. The Korkalovaara suburb continued to grow by less expensive means with concrete element buildings because of a lack of financing and demand (Lukkarinen, 1998).

Figure 2.1 View over Korkalovaara neighbourhood (Photo: Antti Pakkanen).

Currently, Korkalovaara is home to almost 6,000 Rovaniemi residents, and it is one of the most densely populated areas in the city. Despite its somewhat controversial reputation, Korkalovaara has remained a popular place for living since the housing is affordable and is close to daily services and nature.

Even though Aalto's plan was not fully realized in Rovaniemi, the suburban neighbourhoods nationally are a result of planning ideals from an era where they were seen as both a way of modern urban development and a solution for housing issues. In the beginning, these areas were mainly inhabited by working-class families with children. However, societal developments, such as the heavy recession of the 1990s, especially affected the working classes and thus unemployment and social problems increased heavily in the suburban neighbourhoods nationally (Wallin, 2023: 4). As a result, the discussions concerning them began to turn negative. Forest suburbs, which were once seen as idealistic places for living, were labelled as places with criminal and troubled behaviour. The territorial stigma was – and still is – maintained and reinforced by media and public discussions. It has to be noted, though, that the discussion concerning the suburbs has been polarized throughout history. This means that the residents often see them for the most part as good places to live, but from the outside they are seen as undesirable areas (e.g. Roivainen, 1999; Urban, 2011; Maïche, 2021). As Wallin (2023: 4) notes, such territorial stigmatization is also about symbolic power struggles on hierarchies between neighbourhoods and their residents.

In recent decades, the development paths of the Finnish suburbs have been rather negative. As said, the once modern, desirable and future-focused places have gradually been downgraded as areas where undesirable and poor social reputations and positions are present. This is leading to a slow erosion of the places since, along with their reputations, services have also decreased, which in turn has influenced the decline of the areas. In Korkalovaara too, this development has been tangible. In recent years, development in the area has stalled: services have eroded, and house prices are sluggish and rather low. Both public and private services have decreased. One of the culminating points of the downfall of the neighbourhood has been trouble with the local school, which has been one of the many schools in Finland that have been forced to shut down and move because of poor indoor air quality. At the time of writing this text, the school had been demolished for seven years, but a new one is expected to be opened in 2025 after a long process of activist lobbying, negotiation and planning between residents and city officials. Because of the current lack of services, many families have left the area. All these changes make the development streams concerning suburban neighbourhoods nationally visible in Korkalovaara as well (see Stjernberg, 2019). That is, from being a modern residential area of the future to becoming an almost undesirable place to live. At the national level, various development projects and interventions have been initiated to tackle the undesirable development and segregation of these neighbourhoods (e.g. Ympäristöministeriö, 2020).

From the outside, by institutions and persons of power, suburban neighbourhoods are often considered an example of failed planning (Wallin, 2023: 5). Such ideas, along with the territorial stigma attached to these places, construct neighbourhood identities that are hard to change (Wallin, 2023: 4–5). This kind of dissonant heritage is, according to Belli (2020: 87), partly a consequence of class dissonance and segregation. Even though the Korkalovaara neighbourhood is very small compared to the suburbs in large European cities, there are similarities in the dissonance and ambivalence in the stories of these areas. Junnilainen (2020: 45) argues that stigmatized low-income neighbourhoods can be seen not only in a negative light but also as historically and culturally diverse milieus. Junnilainen (2020: 46) points out that the residents who live in stigmatized areas feel belonging and attachment towards their neighbourhoods, which goes against general assumptions (see also Wallin, 2023). She argues that stigmatization as an everyday experience can take the form of non-recognition. Furthermore, based on her research findings, she suggests that neighbourhood-based narratives can serve as resources for collectively employed de-stigmatization strategies (Junnilainen, 2020: 46).

Stories of Korkalovaara: Collecting suburban heritage

As the two previous sections describe, there are many ambivalences in the relationship between tourism and suburban heritage in Rovaniemi and more widely in Lapland. We consider if these could be somewhat resolved and/or overcome through the academic thought and empirical practice of socially sustainable and inclusive tourism (Murphy, 1985; Scheyvens, 2007; Minnaert et al., 2011; Haanpää et al., 2018; Scheyvens and Biddulph, 2018). By taking this stance, we align with the idea that local residents should have the power and possibility to define how tourism impacts the local operating environment (Dredge and Hales, 2012; Timothy, 2012; Höckert, 2015; Hakkarainen, 2017; Paiva, 2023). In other words, our aim as residents of the neighbourhood in question has been to explore if the locality and this untypical heritage could offer alternatives to the local 'tourism norm' and whether such 'counterproductive' tourism production could provide positive impacts for the place and the people living there.

It is certain that suburban neighbourhoods and their heritage are usually not a part of the tourism offerings of cities. Not even if they reveal inspiring stories and noteworthy heritage. Our empirical context, the suburban neighbourhood of Korkalovaara in Rovaniemi, Finland, could be described as an urban periphery. In its material and socio-economical characteristics it differs from the city centre. But, in the city's policies, it has not been recognized as an area with its own identity. As residents, neighbourhood activists and tourism researchers we began to ponder if and how, through tourism activity, the stigma or the non-recognition of the neighbourhood, which Junnilainen (2020, also Wallin, 2023) discusses, could be tackled. How could the neighbourhood and its heritage be recognized, and could these narratives act towards de-stigmatization (Junnilainen, 2020: 46)? As tourism scholars Scheyvens and Biddulph (2018: 593) point out, one of the elements of inclusive tourism is facilitating self-representations by those who are marginalized or oppressed. This would, according to them, allow their stories to be told, and their culture to be represented in ways that are meaningful to them (Scheyvens and Biddulph, 2018: 593). However, 'business-as-usual' tourism shapes places in the city: it often accelerates gentrification and makes places similar and 'easy to consume' (see Paiva, 2023). Perhaps such 'non-interesting' places as suburban neighbourhoods allow us to consider tourism in terms of sustainability: could tourism be done somehow differently and what else could tourism be than what is 'the norm' in Rovaniemi today?

To advance these considerations, we applied and received funding for the 'Stories from Korkalovaara' project, which gathered stories and experiences, namely the intangible heritage, of the residents of the Korkalovaara neighbourhood. These were collected from various time periods and groups of people through different methods. The aim of the project was firstly to build an understanding of the meaning of the neighbourhood to current and former residents as well as part of Rovaniemi city in general. We also aimed to strengthen the community of Korkalovaara and possibly even the vitality of the place. The stories collected in the project were edited and placed on a digital platform (www.tarinoitakorkalovaarasta.fi). Based on the stories, we organized a small-scale event at the end of the project. The aim of this was to test how tourism or touristic practices could contribute to making untypical heritage visible and be a way of learning. The project was in action between 2020 and 2023. As it was established in the wake of the Covid-19 pandemic, it affected the project schedule and also the methods used.

The collection of stories happened through collaboration with the local school, organizing 'storyworkshops' with the local library and as personal interviews. Once the digital platform was published, there was also an opportunity to leave written stories on the internet. These methods of story collection were organized and informed to the public in different ways. To get the stories of children, we established a collaboration with the local school. One grade level, altogether about sixty children, and their teachers participated in a project-based learning module where we first carried out a walking tour with the group to the neighbourhood and told them historical stories about the place as well as factual information about the development of suburban neighbourhoods in Finland. Afterwards, the children worked with their teachers writing their own stories about the neighbourhood and also making board and digital games about the place.

Besides the school collaboration, we organized 'storyworkshops' together with the local library. The library is a very popular place among the residents, especially the senior citizens. The storyworkshops were organized in the style of group discussions. Besides the library advertising them, they were also advertised in the local newspaper. Some people contacted us and stated that they did not want to participate in a group gathering but would like to tell their stories. In the workshops some participants brought material, mainly photographs but also old newspaper articles, with them. These acted as starting points for memories and stories. The participants also added to each other's stories and discussed their memories; how everyone personally remembered certain events, places or happenings. The participants in the workshops were all middle-aged or older, and both workshops lasted about one hour.

Personal interviews were also carried out with three people. These took place in different locations and one of them was realized as a 'driving interview', where we went to see different places of significance by car with the interviewee. The interviews lasted from forty-five minutes to two hours. Afterwards, the interviews as well as the storyworkshops were transcribed verbatim. The stories produced by the school children were already in a written format. We then edited the stories from the material to create shorter versions to place them on the map in the digital platform. The different data resulted in more than 120 short stories and memories. These were from different decades ranging from the 1960s, when the neighbourhood was built, to the 2010s. Once the platform was published, some tens of stories were received through it and it is open for people to leave new entries.

The aim of the 'Stories from Korkalovaara' project was first to collect and gain recognition for the heritage of the suburban neighbourhood. Secondly, it was to consider if and how this could be done through touristic practices. At the end of the project, we organized a two-day neighbourhood event, which consisted of an event publishing the digital platform, walking tours based on the collected stories, art workshops on storying the place and a future oriented workshop of the neighbourhood. These were advertised mainly to the residents of Rovaniemi in the newspapers and different social media channels, but we also distributed advertisements to a few places popular with tourists in the Rovaniemi city centre, for example a local hostel. The event gathered altogether some sixty people, and the most popular events were the publishing of the platform and the walking tours. From the project's point of view, the meaning of the event was to test the means of touristic practices in representing the ambivalent heritage of the suburb. We next move on to discuss what kind of ambivalences were present during the different stages of the project, as well as how these were present in our own understanding of the purpose of the project.

Discussion: Tourism in the suburb?

The ambivalent relationship between suburban heritage and local tourism was very much present throughout the course of the project. We were continually asked by different people and media to explain what these two had in common. 'What is there to see?', was a common question. It was also suggested that we conduct tours based on well-known cultural figures who had lived in the neighbourhood. The stories of ordinary places and people were not thought of as interesting enough by some people. The collection of the stories also counted for a significant proportion of the project time, since these stories had not been seen as valuable before and they did not 'exist' publicly. Besides making an unseen place visible, our aim was namely to bring different histories and places into tourism discussion in Rovaniemi (see Scheyvens and Biddulph, 2018: 590–2). When doing this, it became very clear to us that using such heritage as tourism content in Rovaniemi somehow questioned and contested the normative way of doing tourism, which is much based on a Western consumer society perspective (see e.g. Paiva, 2023). The value system of tourism development and what is considered worthy was very present. Tourism activity in Rovaniemi is not thought of as a phenomenon that can force change or counter the status quo (Scheyvens and Biddulph, 2018: 590–2) but rather as an activity reinforcing them.

Nevertheless, while trying to conduct inclusive tourism practices among the research group, we were continually aware of and frequently discussed the possible ambivalence of our endeavours. The underlining thought of the project was, according to Junnilainen's (2020: 46) argument, to investigate whether neighbourhood-based narratives could serve as resources for de-stigmatization strategies. But, such inclusive practices in tourism development can be ambivalent, because they can in fact enhance the dynamics that they are trying to deconstruct. In other words, the practices intended to be inclusive can add to the sense of exclusiveness. Inclusionary practices, no matter how well-intended they might be, have ambivalence and paradoxes that need to be considered (see Adamson et al., 2021; Kangas-Müller et al., 2023: 5). In the context of inclusive tourism and suburban heritage, this can be seen in Scheyvens' and Biddulph's (2018: 599–600) description of the uncertainty concerning tourism programmes in 'undesirable or unwanted' neighbourhoods, which need to be conducted carefully in order not to commercialize poverty or exploit the marginalized areas in ways that do not benefit them. We did not experience such developments during the course of the project, probably because we were only testing the touristic practices on a small scale. However, if these were to be commercial products sold for wider audiences, these dynamics would have to be considered more carefully.

By organizing the walking tours, besides the alternative narratives, we aimed to test if 'the map of tourism could be reshaped' by organizing touristic activities in an urban periphery (see Scheyvens and Biddulph, 2018: 590). As Scheyvens and Biddulph (2018: 590–2) argue, such tourism activity can act on the one hand as a force of change by addressing inequalities and segregation of different places and groups of people, and on the other hand by challenging stereotypes and conventional interpretations of history. Indeed, the walking tours were a powerful method in the project (we also organized a few of them outside the project actions) to raise awareness and share information on the unconventional heritage of the suburban neighbourhood (see Haanpää et al., 2022). In the touristic practice of walking, people's senses become heightened, and they observe their surroundings more carefully (Österlund-Pötzsch, 2010). As Carú and Cova (2003) argue, engaging with the flows of everyday life by walking provides premises for experiences. However, the form of an organized walking tour is also a way of representing the place where it is held.

A touristic walking tour can be understood as a form of storytelling that is a poetic and political practice (Lund, 2008; see also Aoki and Yoshimitzu, 2015:

Figure 2.2 Walking tour participants in front of the neighbourhood mall (Photo: Antti Pakkanen).

276). The narration of the tour gives meaning and value to places encountered and shapes the participants' understanding of them. As such, we had to carefully consider our ways of telling the stories of the neighbourhood. However, the tours often created a certain atmosphere between us guides and the participants, where different stories were mutually shared. It was not just the 'official' story told by us guides, but also the participants engaged in storytelling. Often, local, random by-passers participated in the storytelling too. They were first interested in what was going on as a group of people on a touristic walk is not a typical sight in the neighbourhood. Once they were told what we were doing, many were suspicious of our purposes but still wanted to share their own stories. It has to be noted though that, even if the by-passers were curious about our doings, their attitude might change if such an activity happened every day in the neighbourhood.

We also had to consider the question of representation (e.g. Alvarado-Sizzo, 2023). This manifested in both collecting the stories and publishing them in digital form but also (re-)telling them on the walking tours. The project resources affected our possibilities to collect stories and this influenced the people who participated in the study. Also, the ways the data was collected somewhat reinforced the fact that more people with middle-class backgrounds shared their stories (see e.g. Luhtakallio and Mustranta, 2017). This naturally shaped the way the heritage of the neighbourhood is presented: it is certain stories that are told, and others go unheard. The publication of the stories on the digital platform added 'another layer' to the question of representation, since the original transcriptions were edited by us, the researchers. This was a research ethical question of anonymity, but the chosen way of presenting the stories collected on a map forced us to leave out some stories if the place on the map connected with the story would be too recognizable. We were also forced to edit and shorten some stories quite radically for them to fit the chosen form. Through the platform we also received stories that were ambivalent in their content: they described drug use and other social problems in detail. As we did not want to reinforce territorial stigma, we were forced to contemplate how to edit and publish these stories. We tried to navigate these difficulties by discussing the decisions together with the research group (the authors of this chapter) and being self-reflexive towards our choices.

Conclusion

This chapter has presented and discussed viewpoints on ambivalent heritage and tourism in the Korkalovaara neighbourhood in the city of Rovaniemi, a popular tourist destination in Finnish Lapland. Korkalovaara is not typically mentioned as part of the tourism map of Rovaniemi. On the contrary, it has been seen as an unwanted neighbourhood, much like other similar suburban neighbourhoods in Finland and other parts of Europe.

First, we have argued that the heritage in places such as the Korkalovaara suburban neighbourhood is a valuable resource, but also has an ambivalent component. This ambivalence is rooted in the experiences of the locals and in the past developments in the neighbourhoods. The neighbourhoods are seen on one hand as a good place to live and places of living heritage by the residents, but, on the other hand, they are seen – mostly from the outside – as undesired or unwanted places, with persistent stigma attached to them and the residents. In a forest suburb, the closeness of nature and possibilities to enjoy outdoor activities are present, which can be seen as valuable resources. However, this kind of geographical position also means that the suburban neighbourhoods are usually located on the city outskirts, which creates a space for social segregation. Thus, the suburban neighbourhoods can be seen as places of contradictions and dichotomies.

The ambivalence of the place and its heritage is also rooted in the means and meanings of tourism development, which might be exclusive in nature or enhance the sense of exclusion for those they are trying to include. Tourism in Rovaniemi is mainly developed on the basis of winter tourism and Santa Claus.

Although tourism development has been successful in Rovaniemi and the economic impacts are significant, the role of locals and local heritage, especially that of the suburban neighbourhoods, has been in the margins. Consequently, the contrast between wanted and unwanted places grows when the undesirable neighbourhoods are discussed in the tourism context. In the project, we tested tourism practices that are based on the unseen, and as such ambivalent, heritage of the suburb. These were rather moderate compared to the amplitude of tourism activity in Rovaniemi. However, several walking tours were created and executed during the years. These proved to be of interest, at least for certain audiences. Additionally, from the comments of the participants we could interpret that the tours were important for them, either because of learning about the neighbourhood and its history and/or because of sharing their own experiences.

Scheyvens and Biddulph (2018: 601) argue that, along with the economic benefits, the social benefits of tourism are important, even though they do not receive as much attention as the economic ones. According to them, inclusive tourism has the potential to result in enhanced unity in rural and urban locations and a more pleasant place to live (Scheyvens and Biddulph, 2018: 601). However, it is important to consider how tourism is developed in these places: for whom and on whose terms. Not all will want to be part of tourism or see the neighbourhood as a touristic place, which makes the issue complicated. Accordingly, it can be argued that this kind of tourism is not for the masses, since the value aimed for is rather more social than economic. It may be more relevant for individual travellers or small groups. The experience is then based on the local life as it is, storytelling and nostalgia. Small-scale touristic activities, such as walking tours, have gained attention, and participants in Korkalovaara have shown interest in telling or sharing stories from their past. It seems though, that while the residents or former residents are willing to participate and tell stories, it has to happen on their own terms. Tourism has the potential to create social value in suburban neighbourhoods, but there are many aspects to consider in its development that can only be resolved through careful listening and local solutions.

References

Aalto, P. and M. Laaksonen (2021), *Turun lähiöt: Aluerakentaminen Turun kaupunkiseudulla*, Turku: Sammakko. [The suburban neighbourhoods of Turku.]

Adamson, M., E. Kelan, P. Lewis, M. Śliwa and N. Rumens (2021), 'Introduction: Critically interrogating inclusion in organisations', *Organization*, 28 (2): 211–27.

Alvarado-Sizzo, I. (2023), 'Spatial representations, heritage and territorial-synecdoche in contemporary tourism', *Tourism Geographies*, 25 (2–3): 467–86.

Aoki, J. and A. Yoshimitzu (2015), 'Walking Histories, Un/making Places: Walking Tours as Ethnography of Place', *Space and Culture*, 18 (3): 273–84.

Belli, N. (2020), 'Modern Urban Planning and Dissonant Heritage: The Case of San Polo', *Art History & Criticism / Meno istorija ir kritika*, 16 (1): 79–93.

Berkenbosch, K., P. Groote and A. Stoffelen (2022), 'Industrial heritage in tourism marketing: Legitimizing post-industrial development strategies of the Ruhr Region, Germany', *Journal of Heritage Tourism*, 17 (3): 327–41.

Carù, A. and B. Cova (2003), 'Revisiting Consumption Experience: A More Humble but Complete View of the Concept', *Marketing Theory*, 3 (2): 267–86.

Chadha, A. (2006), 'Ambivalent Heritage. Between Affect and Ideology in a Colonial Cemetery', *Journal of Material Culture*, 11 (3): 339–63.

Donaldson, E. (2018), 'Place, Destabilized: Ambivalent Heritage, Community and Colonialism in the Marquesas Islands', *Oceania*, 88 (1): 69–89.

Dredge, D. and R. Hales (2012), 'Community case study research', in L. Dwyer, A. Gill and N. Seetaram (eds), *Handbook of Research Methods in Tourism: Quantitative and Qualitative Approaches*, 417–37, Cheltenham: Edward Elgar Publishing.

Euromonitor (2014), 'World Travel Market Global Trends Report 2014', available online: https://www.globalwellnesssummit.com/wp-content/uploads/Industry-Research/Global/2014-wtm-global-trends-report.pdf (accessed 5 August 2023).

Gonzalez-Soria de la Santa, J. (2020), 'Tourism as a lever for a more inclusive society', *Worldwide Hospitality and Tourism Themes*, 12 (6): 731–8.

Haanpää, M., J.-C. García-Rosell and M. Hakkarainen (2022), 'Walking the concepts: elaborating on the non-representational sensitivities of tourism experience', *Journal of Marketing Management*, 38 (15–16): 1832–50.

Haanpää, M., M. Hakkarainen and A. Harju-Myllyaho (2018), 'Katsaus yhteiskunnalliseen yrittäjyyteen matkailussa: osallisuuden mahdollisuudet pohjoisen urbaaneissa paikallisyhteisöissä', *Matkailututkimus*, 14 (2): 44–58. [A review on social entrepreneurship in tourism.]

Hakkarainen, M. (2017), *Matkailutyön ehdot syrjäisessä kylässä*, PhD dissertation, Acta Universitatis Lapponiensis 357, Rovaniemi: University of Lapland. [The terms of tourism work in a remote village.]

Hakkarainen, M. and T. Leskinen (2001), *Kylien tarinat-elämää Kurtakon, Venejärven ja Teurajärven kylissä,* Rovaniemi: University of Lapland. [Stories of the villages.]

Hakkarainen, M. and P. Satokangas (2022), 'Tupa tunturin laidalla', *Kleio*, 4: 20–1. [A Cottage on the edge of the Fells.]

Herva, V.-P., A. Varnajot and A. Pashkevich (2020), 'Bad Santa: cultural heritage, mystification of the Arctic, and tourism as an extractive industry', *The Polar Journal*, 10 (2): 375–96.

Höckert, E. (2015), *Ethics of Hospitality: Participatory Tourism Encounters in the Northern Highlands of Nicaragua*, PhD dissertation, Acta Universitatis Lapponiensis 312, Rovaniemi: University of Lapland.

Ilola, H., M. Hakkarainen and J.-C. García-Rosell (2014), 'Johdanto Rovaniemen joulumatkailuun', in H. Ilola, M. Hakkarainen and J.-C. García-Rosell (eds), *Joulu ainainen*, 10–22, Rovaniemi: Multidimensional Tourism Institute. [Introduction to Christmas tourism in Rovaniemi.]

Järvinen, A. E. (1932), *Suurissa metsissä: Eränkäyntejä Lapissa ja Perä-Pohjolassa,* Porvoo: WSOY. [In the large forests. Wilderness acts in Lapland and Perä-Pohjola.]

Junnilainen, L. (2020), 'Place Narratives and the Experience of Class: Comparing Collective Destigmatization Strategies in Two Social Housing Neighborhoods', *Social Inclusion*, 8 (1): 44–54.

Kangas-Müller, L., K. Eräranta and J. Moisander (2023), 'Doing inclusion as counter-conduct: Navigating the paradoxes of organizing for refugee and migrant inclusion', *Human Relations,* online ahead of print.

Komppula, R., S. Hakulinen and S. Saraniemi (2009), 'Lapin joulucharter-tuotteen elinkaari', *Matkailututkimus*, 5 (2): 19–42. [The lifecycle model of the Lapland Christmas charter product.]

Kortteinen, M. (1982), *Lähiö: Tutkimus elämäntapojen muutoksesta*, Keuruu: Otava. [Suburban neighbourhood: A study on changing ways of life.]

Koskinen-Koivisto, E. (2023), 'Contested and Ambivalent Heritage: Revisiting Responses to the Second World War Heritage of Finnish Lapland from the Perspective of Affects', in C. Axelsson Yngvéus, M. Thor Tureby and C. Trenter (eds), *(Un)contested Heritage: Archives, Museums and Public Spaces*, 175–89, Malmö: Malmö Universitet.

Luhtakallio, E. and M. Mustranta (2017), *Demokratia suomalaisessa lähiössä*, Helsinki: Into. [Democracy in a Finnish suburban neighbourhood.]

Lukkarinen, P. (1998), *Aalto Lapissa: Alvar Aallon Lapin tuotanto*, Jyväskylä: Atena. [The works of Alvar Aalto in Lapland.]

Lund, K. A. (2008), 'Listen to the Sound of Time: Walking with Saints in an Andalusian Village', in T. Ingold and J. L. Vergunst (eds), *Ways of Walking: Ethnography and Practice on Foot*, 93–103, Aldershot: Ashgate.

Maïche, K. (2021), 'Puhetta lähiöstä', in E. Puumala, K. Maïche and S. Tuomaala-Özdemir (eds), *Naapurusto – Tarinoita moniäänisestä lähiöstä*, 20–37, Helsinki: Into. [Talk about suburban neighbourhood.]

Martini, A. and D. M. Buda (2020), 'Dark tourism and affect: Framing places of death and disaster', *Current Issues in Tourism*, 23 (6): 679–92.

Minnaert, L., R. Maitland and G. Miller (2011), 'What is social tourism?', *Current Issues in Tourism*, 14 (5): 403–15.

Murphy, P. E. (1985), *Tourism: A Community Approach*, London: Routledge.

Österlund-Pötzsch, S. (2010), 'Pedestrian Art', *Ethnologia Europaea*, 40 (2): 14–28.

Paiva, D. (2023), 'The paradox of atmosphere: Tourism, heritage, and urban liveability', *Annals of Tourism Research*, online ahead of print: 1–10.

Paulaharju, S. (1963), *Wanhaa Lappia ja Peräpohjolaa*, 2nd edition, Porvoo: Werner Söderström. [Old Lapland and northern Ostrobothnia.]

Paulauskaite, D., R. Powell, J. Coca-Stefaniak and A. M. Morrison (2017), 'Living Like a Local: Authentic Tourism Experiences and the Sharing Economy', *International Journal of Tourism Research*, 19 (1): 619–28.

Roivainen, I. (1999), *Sokeripala metsän keskellä – lähiö sanomalehden konstruktiona*, Helsinki: Helsingin kaupungin tietokeskus. [Suburban neighbourhood as a newspaper construction.]

Saari, R., E. Höckert, O. Kugapi, M. Lüthje and N. Mazzullo (2020), 'Cultural Sensitivity in Sámi Tourism: A Systematic Literature Review in the Finnish Context', *Matkailututkimus*, 16 (1): 93–110.

Scheyvens, R. (2007), 'Exploring the Tourism-Poverty Nexus', *Current Issues in Tourism*, 10 (2): 231–54.

Scheyvens, R. and R. Biddulph (2018), 'Inclusive Tourism Development', *Tourism Geographies*, 20 (4): 589–609.

Stjernberg, M. (2019), *Concrete Suburbia: Suburban housing estates and socio-spatial differentiation in Finland*, PhD dissertation, The Department of Geosciences and Geography, University of Helsinki, A 77.

Timothy, D. J. (2012), 'Destination communities and responsible tourism', in D. Leslie (ed.), *Responsible Tourism: Concepts, Theory and Practice*, 72–81, Wallingford: CABI.

Totto (2023), Yhdistyksen historiaa [History of the association]. https://www.totto.fi/kotiseutuyhdistys-rovaniemen-tot/yhdistyksen-historiaa/. Retrieved on 15 December 2023.

Urban, F. (2011), 'Mass Housing in Germany – Controversial Success and Ambivalent Heritage', in J. Lizardi and M. Schwegmann (eds), *Espacios Ambivalentes: historias y olvidos en la arquitectura social moderna*, 52–75, San Juan, Puerto Rico: Ediciones Callejón.

Visitory.io (2023), Rovaniemi. Majoitustilastot ja Matkailutilastot. [Accommodation and Tourism Statistics.] https://visitory.io/fi/public/rovaniemi/2022-12/. Retrieved on 15 December 2023.

Wallin, A. (2023), 'Green Neighbourhood Identity: How Residents Use Urban Nature Against Territorial Stigmatization in Finnish Housing Estates', *Housing, Theory and Society*, online ahead of print, 1–19.

Wallin, A. and S. Rosendahl (2017), 'Kerrostalolähiön kulttuurisen muutoksen kartoitus', in M. Häyrynen and A. Wallin (eds), *Kulttuurisuunnittelu: Kaupunkikehittämisen uusi näkökulma*, 71–102, Helsinki: Suomalaisen kirjallisuuden seura. [A study on cultural change of a suburban neighbourhood.]

Ympäristöministeriö (2020), *'Lähiöohjelma 2020–2022: Hyvinvoivat vetovoimaiset lähiöt'*. Available online: https://ym.fi/documents/1410903/38439968/Lahiohjelma_2020-2022-3B1828A2_FB2B_4C4D_9D13_BE21A27F26F3-155188.pdf/44f93e8a-3f53-dfb4-39a0-e08f08b8c70b/Lahiohjelma_2020-2022-3B1828A2_FB2B_4C4D_9D13_BE21A27F26F3-155188.pdf?t=1603259591934 (accessed 7 August 2023). [The national development programme for suburban neighbourhoods 2020–2022.]

3

Local Activism and Agency in Heritagization Processes: The Case of the Forests at the Paimio Sanatorium, Finland

Maija Mäki and Helena Ruotsala

Introduction

The Paimio Sanatorium is a world-famous architectural site, but in Finland it is still not so well-known, even though its architectural significance has long been acknowledged. The Paimio Sanatorium was designed by the Finnish architects Alvar and Aino Aalto. It was opened in 1933 and served as a tuberculosis sanatorium until the early 1960s. It was then used as a general hospital until 2015. Currently, the sanatorium is being transformed into a cultural centre and heritage site by the Paimio Sanatorium Foundation. There are plans to nominate the sanatorium, along with twelve other buildings designed by Alvar Aalto, as UNESCO World Heritage sites. The role played by locals has been significant from the beginning of the sanatorium's life cycle, ever since construction of the site. The different periods of the sanatorium and the roles of the locals emerge in the oral history materials that have been collected during the past few years.

In addition to the hospital and staff buildings, Paimio Sanatorium owned extensive forest areas totalling more than 400 hectares. In 2018, the forests in the sanatorium area were sold to a private owner, who started cutting them down. A group of locals became angered by the disappearance of the forests, calling attention to their importance as an essential part of the sanatorium complex. The pine forests had been an integral part of the tuberculosis sanatorium because of the believed health benefits of the trees. For example, patients took regular walks in the forests and some of them also engaged in 'forbidden walks' and secret romances in the woods with other patients and also sometimes with staff members.

This chapter focuses on the history of the Paimio Sanatorium's forests. A multidisciplinary project entitled 'Paimio Sanatorium: Social, Historical and Cultural Perspectives', by scholars from the fields of ethnology, folklore and museum studies at the University of Turku, focused on collecting memories and stories connected to sanatorium forests and local opinions concerning the future of the sanatorium and its environment. The role of the local activists has been an important part of the research project, especially with respect to the heritagization process of the environment and forests of the sanatorium. In this chapter, we examine the processes of heritagization and the meanings and values that locals have assigned to the forests at different times. Our main questions are as follows: What characterizes the agency of local-level activism, and how has it been affected on a practical level in the case of the Paimio Sanatorium? We are also interested in what factors led the locals to become active in protecting it as a heritage site and what kind of role they have played in the heritagization process of the pine forests surrounding the sanatorium.

The theoretical framework of the study arises from the fields of critical heritage studies and environmental ethnology. The importance of environment has already been considered in many prior studies in Finnish ethnology (e.g. Ruotsala, 2002; Korjonen-Kuusipuro, 2012; Sonck-Rautio, 2019). People have long used the resources of their respective environments to their advantage, whether by cutting down forests, utilizing heather or other natural resources, or selling berries, to name but a few. Studies focusing on how people learn to make use of their natural surroundings fall into a category called environmental ethnology. Anthropologists Nora Haenn and David Casagrande (2007: 101) have asserted that anthropologists and ethnologists work as cultural brokers on environmental issues since their skills as delineators and negotiators when working with different stakeholder and identity groups benefit policymakers. In the field of critical heritage studies, scholars have pointed out a need for ethnological perspectives and ethnographic fieldwork to understand how the relationships between local actors, heritage professionals and bureaucrats and the objects, places and practices of heritage, including global institutions, are formed and reformed through debates and actions (Smith, 2007; Harrison, 2013).

The research material for the chapter consists of interviews and existing literature. The interviews were conducted at the beginning of 2023 and are now kept in the Archives of the School of History, Culture and Arts Studies (SHCAS) at the University of Turku (TYKL/aud/1848–1851). They include four interview appointments and six interviewees. The interviewees discussed here have been anonymized. If necessary, we use the professional names of the interviewees as pseudonyms ('Teacher', 'Local Politician', 'Forester', 'Lawyer', 'Entrepreneur' and 'Architect'), which is intended to highlight their expertise in the process of preserving the forests of the Paimio Sanatorium. It must be noted that the topic of the article is sensitive in several ways. The tuberculosis sanatorium itself carries the weighty heritage of a medical burden, meaning sanatoriums are controversial from a contemporary standpoint. But, viewed from another angle, the future of the sanatorium has evoked strong opinions and feelings in local politics over the past few years (e.g. Salo, 2018; Tukia, 2019).

Research concerning the Paimio Sanatorium has often contained a strong architectural emphasis (e.g. Heikinheimo, 2016; Heimo, 2022). Ethnological research, though, focuses on the everyday life of the sanatorium and people's experiences and memories. Their individual experiences are usually constructed bodily and create affective knowledge of the environment of the sanatorium (e.g. Jäntti et al., 2022). This study can thus complement and deepen current understandings of the agency of the local people connected to heritagization processes in general and specifically the significance attached to the Paimio Sanatorium and its environment at different times.

Local activism as an act of memory activism

In ethnology and anthropology, various perspectives concerning the use of the environment and the changes and threats resulting from such use have gradually piqued the interests of researchers. Studies have focused on, for example, how the environment and its resources have been exploited or how the people living in them have adapted to changes (Ruotsala, 2010). In Finland, several previous protection battles have been fought, especially over rivers and hydropower production in the north (e.g. Suopajärvi, 2001; Ruotsala, 2013; Nieminen, 2022). The people behind the protection movements were local actors. Leena Suopajärvi (2001), who has examined the struggles over access to rivers, saw the differences in the argumentation made by locals and those representing the hydropower economy. The local argumentation was driven by daily life experiences and the vernacular heritage connected to the river.

In recent years, discussions about vernacular heritage and memory activism have made visible the fact that ordinary people are active agents in the creation, interpretation and utilization of their own history and heritage (e.g. Ashton and Kean, 2008; Robertson, 2016; Bakalarz-Duverger, 2023). Vernacular heritage addresses the ordinary and everyday aspects of the past and present that are challenging conventional authorized heritage discourses (Robertson, 2016: 6–8). Vernacular heritage practices open a space for understanding and interpreting the expressions and performances of heritage produced by local people. Dynamic and powerful expressions of such practices might provoke social changes, as has happened in environmental battles before. Based on this background discussion, we understand and emphasize the ambivalent nature of heritage especially from the perspective of the 'below' (e.g. Robertson, 2016).

In the case of the Paimio Sanatorium, it is possible to follow the agency of locals from past to present. Locals have experienced the different stages of the sanatorium and some even worked at the sanatorium when it was first constructed in the early 1930s. The different levels of hierarchies can be traced over time, and the agency of locals has also changed throughout the past few centuries. During the process of protecting the sanatorium's forests, ongoing negotiations took place regarding the power of different agencies. Heritage authorities, city bureaucrats, destination management workers, locals and finally also researchers have constantly had discussions about what kinds of roles different actors can or should play. Locals saw the importance of heritage authorities and researchers for strengthening the cultural heritage status of the forests. Although the significance of the pine forests surrounding the sanatorium was clear to professionals, locals were the ones who became activists. The agency of locals can therefore be seen through the lens of memory activism. Cultural heritage evokes emotions and is quite affective with respect to nature. Activism, directed at remembering, protecting and preserving cultural heritage sites, highlights the specific emotional charge that the concept of ambivalent heritage seeks to achieve. What makes the Paimio Sanatorium a particularly interesting case is how its architectural significance is intertwined with its surroundings, especially the forest, and how specifically the fate of the forest stimulated local agency, prompting people to become activists.

A brief history of the Paimio Sanatorium and tuberculosis in Finland

The Paimio Sanatorium is often referred to as the most internationally well-known building in Finland (e.g. Heimo, 2022). It was opened in 1933 as a regional hospital for tuberculosis patients. It was designed by the young Finnish architects Aino and Alvar Aalto. They designed it as 'a healing instrument', and today the sanatorium is considered the principal work from their early functionalist period.

The decision was made in 1928 to build the sanatorium in Paimio. One option was 40 hectares of land owned by the local Spurila manor. In a statement issued during the decision-making process, the experts described the area and its suitability for sanatorium use as follows:

> In that area, it is easy to find a place wide enough, flat enough and [facing] the south, where a sanatorium can be placed. There is probably enough water available and a sewage plant would obviously be easy to arrange. The railway station is 3–4 kilometres away. And a wide range of building needs are available around the locality. In addition, Paimio is an old, civilized area that both the patients and the staff of the sanatorium would likely enjoy.
>
> Aalto et al., 1933: 9

The building process lasted five years, from 1929 to 1933 (Figure 3.1). As with other tuberculosis sanatoria throughout Finland, the Paimio Sanatorium was built in the middle of a young pine forest where the air was perceived to be fresh, dry and clean (Finnish Heritage Agency, 2009; Hakosalo, 2020). The models for

Figure 3.1 The newly opened Paimio Sanatorium (Photo: Gustaf Welin 1933, Alvar Aalto Museum).

the sanatoria were obtained from Germany, where treatment was based on fresh air, good nutrition, rest and appropriate exercise to improve the patients' general condition and thus increase their natural resistance to disease (Saarikko, 2007: 19). These same principles were adopted in Finland. The first patients came to the sanatorium in February 1933. After two months, all the hospital's available rooms were occupied by patients (Törrönen, 1983: 48). The history of the Paimio Sanatorium can, from that point, be divided into three distinct periods. During the Second World War, it served as temporary housing for the displaced Karelian population (Jaakkola, 2015). Then, it served as a tuberculosis sanatorium until the early 1960s, and thereafter as a general hospital until 2015 (see, e.g. Finnish Heritage Agency, 2009; Heikinheimo, 2016). The decision to close the hospital was made in 2012, with all hospital operations ending in 2015.

Throughout the twentieth century, tuberculosis was a very dangerous disease, killing tens of thousands of people in Finland. It was only after 1964 that the number of cases began to noticeably decline. What was alarming to society about tuberculosis was that many of those who died from the disease were in their twenties or forties, i.e. in prime working age. Tuberculosis also lasted a long time: patients struggled with the disease for an average of seven years (Tilastokeskus: Tautikuolleisuus, 1936–2010; Savonen, 1936: 208; Pesonen, 1980: 495).

Life in the tuberculosis sanatoria is usually described based on two principal narratives. On the one hand, tuberculosis was a disease that afflicted creative and suffering artists and rich travellers in times past, many of whom spent their time in sanatoriums in Central Europe. This narrative has been popularized in literature, films and plays. On the other hand, a more public narrative focuses on tuberculosis as an inherited affliction. Tuberculosis was at the time linked to people from lower social classes with poor hygiene. The healing periods in sanatoria may have caused much trauma, associated with feelings of shame, guilt and fear, and therefore, those memories have been silenced and not talked about openly in public. Some oral history projects have included specific goals of dispelling the stigma surrounding the illness and increasing people's understanding of everyday life in the sanatoria (e.g. Parantolaperinnetta, 1972; Shaw and Reeves, 2009; Kelly, 2011). For example, the historian Heini Hakosalo (2021) has demonstrated that patients often linked their personal histories to national narratives as a way of giving a sense of purpose and meaning to their personal suffering.

The experiences of the patients themselves and the types of agency available to them in a sanatorium environment that was considered quite hierarchical are topics that have inspired recent research on the Paimio Sanatorium (Havula,

2021; Kokko, forthcoming). Before effective antituberculosis drugs became available in Finland at the beginning of the 1950s, sanatorium periods were typically long. Most of the patients were young people, and they did not just lie in the beds but developed programmes on their own. For instance, they started their own newspaper, called *Pasuuna*, organized parties and other activities, and rebelled against the strict rules of the sanatorium. They experienced romances, not just between other patients, but also with staff members. They also continued with their lives after receiving treatment (Meurling, 2022: 4). The sanatorium communities were relatively closed spaces. In Paimio, the sanatorium was located behind a barbed-wire fence. For example, the memories of the staff's children who lived in the sanatorium area reveal the previously hidden and unknown aspects of everyday life (for more on child patients, see, for example, Shaw and Reeves, 2009; Burke, 2018). Considering such experiences, memoirs and testimonies, the sanatorium not only appears as an architecturally significant object, but also as a site attached to various agencies and assigned various cultural meanings, carried not only by the patients of the sanatorium but also by those who worked or lived there as adults or children over the years (Figure 3.2).

Figure 3.2 Patients lying on the solarium terrace: below is a walking garden with a round water pool and a parkway; the pine woods are young (hardly any photographs of patients in the sanatorium have been published, but this photo is one of the few that shows presumed patients, taken by Gustaf Welin in 1934, Alvar Aalto Museum).

The natural environment surrounding the sanatoria strengthened patient engagement and played an important role in their daily lives, especially for those who were in relatively good health. Based on the collections on sanatorium traditions, Heini Hakosalo (2020: 70) describes 'lumpustelu' (in American sanatorium slang, 'cousining') as basically any female–male interaction, but typically consisting of long walks in the pine forests. This word does not exist in Finnish outside the sanatorium context. For example, there is a pond called 'Lemmenlampi', or 'Love Pond', near the Paimio Sanatorium. The name was quite typical: sanatoria were surrounded by 'kissing alleys' and 'love paths'. We do not know how common intimate relationships were at sanatoria, but contacts between young people and relaxed walks outside the sanatorium walls were probably important for many patients who had to stay in the sanatorium for long periods and who had little contact with the outside world (Hakosalo, 2020: 70). Other forbidden activities also took place in the woods, for example smoking, drinking and playing cards. The activities demonstrate the agency of patients in ways that challenged the hospital authorities (Portelli, 1997: 7–8; Meurling, 2022: 4–6).

After the 2012 decision to close the Paimio Sanatorium, it took years for it to achieve its current heritage status, having initially been listed for public sale after closure of the hospital. The sale ultimately did not take place because the Paimio Sanatorium Foundation assumed control of the site. The foundation was founded in 2020 by the Hospital District of Southwest Finland, the State of Finland, the Alvar Aalto Foundation, the Municipality of Paimio and the City of Turku. The purpose of the foundation is to protect the sanatorium's facilities and movable assets as well as its intellectual and material property (Paimio Sanatorium Foundation, 2023). The sanatorium opened to the public in June 2021, but only the main buildings and their immediate surroundings in the sanatorium area. The process behind the protection of the sanatorium's forests is a different matter.

The process of protecting the forests around the Paimio Sanatorium

> *It feels like it rises from the moor, this whole complex. This is like the most essential subject here.*
>
> 'Architect', TYKL/aud/1851

In the beginning of the 1930s, a proper sanatorium needed its own farm where it could produce local, fresh food and provide opportunities for patients to work.

In Paimio, farms with the names Heikkilä and Valkiapää were purchased for the sanatorium, comprising a total area of 270 hectares, including 60 hectares of cultivated land. The area of the farm increased again in 1963, when more land was purchased for the sanatorium. The sanatorium then included more than 400 hectares of land. As time passed, though, the hospital only used a small part of the land first purchased for the sanatorium. In 1983, most of the land was sold to the State Forest Research Institute and then to the municipality of Paimio, leaving a total of approximately 62 hectares at the hospital's disposal (Saarikko, 2007: 16).

Even though the Paimio Sanatorium buildings were taken off the public market, in the winter of 2018 the forests around the Paimio Sanatorium were sold to a private landowner by the Hospital District of Southwest Finland for commercial forestry purposes and the construction of small houses. Many locals who lived nearby, as well as some who recognized the value of the unique cultural environment, shared the same concern about the fate of the forests.

> For us, of course, the forests were in daily use and the whole environment of the sanatorium became important for that reason.
>
> 'Teacher', TYKL/aud/1848

> I guess we wouldn't live here, not without the forests.
>
> 'Lawyer', TYKL/aud/1849

> At first, I thought it was not worth it, that this [attempt to save the forests] is crazy. But this is my mother's home district, and my old mother is still alive.
>
> 'Local Politician', TYKL/aud/1848

> For me, somehow an inspirational factor is Alvar Aalto himself. The meanings I see in the forests can be spun around . . . this reddish-brown colour and vertical lines of the pine trees that repeat in detail, all those myths and stories and all those great ideas developed around him.
>
> 'Architect', TYKL/aud/1851

The local people who began working voluntarily to preserve the forests mostly did not know each other beforehand, even though many actively followed the changes in ownership of the sanatorium and its forests. The interviewees shared the same concern for the preservation of the forests. Persons labelled 'Teacher', 'Forester' and 'Lawyer' lived near the sanatorium area, while those labelled 'Architect' and 'Local Politician' are locally known persons who have participated in previous protection disputes. The person labelled 'Entrepreneur' was also active in local politics at the time, but in a different political party than the 'Local Politician'. The person labelled 'Forester' had been hired by the municipality of Paimio, but his wife, the 'Lawyer', does not have any connections to politics or the municipal administration. Likewise, the 'Teacher' had not personally known any other persons beforehand. The 'Teacher' and 'Architect' met in the local communal house, situated in a nationally significant meadow area, and started to talk about the situation with the forests. Already before this discussion, the 'Forester' and 'Lawyer' had initiated a public debate on the future of the sanatorium forests. When all these people, except for the 'Entrepreneur', found each other, they decided to set up a petition to collect the names of those opposing the sale of forests to private owners (called 'The petition to acquire Sanatorium forests for the city of Paimio', 17 March 2019). The petition stated that, 'The values of the sanatorium and the different new uses of the site, as well as the possibility of a UNESCO World Heritage Listing (Network of Aalto Cities), can only be secured with sufficiently extensive preservation efforts and the thoughtful maintenance of forests.' The petition was signed by more than 300 people and presented to local politicians at a city board meeting. However, the board opted not to approve the petition, and so the city of Paimio did not acquire the forests.

'I quickly realized that the city is not going to play any role in this [process]', commented the 'Entrepreneur' when explaining his decision not to support the purchase of forests for the city (TYKL/aud/1850). He contacted the 'Teacher' and 'Lawyer' and proposed a project for them to purchase the forests for the locals. He already had one possible donor in mind, but others doubted the success of the project (TYKL/aud/1850; TYKL/aud/1848; TYKL/aud/1849). First, the 'Entrepreneur' called the private landowner and received a positive response for the project. Then, he contacted a local, wealthy family whose members joined the fundraiser. At this point, there was already a renewed sense of hope among the other group members as well. The 'Architect' had contacts with a major private foundation in Finland, the Kone Foundation, and after an intensive discussion it was confirmed that the project would receive the largest donation from the Kone Foundation: 250,000 euros. The Real Estate Foundation also donated a significant amount to the preservation project. At the same time, the 'Entrepreneur' negotiated the final price for the forests with the landowner, and other members of the group began to establish an association for the protection and management of the forests. They also produced project presentation material for potential donors (Figure 3.3) (Jokela and

Figure 3.3 Cover illustration of the presentation material for potential donors, made by the volunteers: 'Come and save the cultural forests of the Paimio Sanatorium' (Parantolan metsät ry 2019).

Pihkala, 2019; TYKL/aud/1848; TYKL/aud/1849; TYKL/aud/1850; TYKL/aud/1851).

When the purchasing price for the forests was announced, they realized that they needed more private donors to supplement the funding. Members of the group called their contacts asking for donations. Finally, the effort resulted in a total of forty donors, including local companies and private individuals both in Paimio and further afield. The total amount of donations exceeded 450,000 euros. Since the situation was quite sensitive locally, the project was carried out in complete secrecy and was not made public until after the transaction had been completed. In practice, the donations were raised in a couple of months. The new non-profit association, Parantolan metsät ry ('The Forests of the Sanatorium Association') acquired the pine forests surrounding the Paimio Sanatorium on 17 September 2019. The total area of forest land purchased was 28 hectares (Jokela and Pihkala, 2019; TYKL/aud/1848; TYKL/aud/1849; TYKL/aud/1850; TYKL/aud/1851).

In a nutshell, the story behind saving the forests of the Paimio Sanatorium is a success story of civic activism, and the result was achieved through the efforts of six individuals. However, the project unfolded in different stages, and the participants' descriptions of the process included a mixture of emotions: disbelief, despair and disappointment combined at the same time with a sense of hopefulness, cohesion and faith in the future.

> I couldn't really believe this [success]. I have followed local politics for quite a long time. I have become terribly cynical and a little bit parochial. But my true nature is not like that, as my wife knows what I am like.
>
> 'Local Politician', TYKL/aud/1848

> I reacted more strongly to it [than my husband]. I lost sleep at night because I worried that all the forests would be cut down so that I wouldn't be able to live here anymore. In a way, I wanted to save the neighbourhood and the children's school and school roads.
>
> 'Lawyer', TYKL/aud/1849

The interviewees described the fundraising process as a roller coaster. For example, even after the money for the forest land had been collected, they still needed to pay the transfer tax. There had initially been talk that the city might contribute to the costs of the transfer tax, but in the end the city did not participate in paying the tax, which was more than 18,000 euros. This decision meant that volunteers had to find donations to cover the taxes as well.

> 'We were really tired. It took so much effort to get so many donations. But then we realized that we had no other choice but to just get up and get the missing money. At this point, you can't be discouraged. We started to call people and turn [over] all the stones and stumps. . . . We started to build up the missing amount from the smaller streams [of money] and finally we succeeded.'
>
> 'Entrepreneur', TYKL/aud/1850

The purpose of The Forests of the Sanatorium Association is to manage the pine forests surrounding the Paimio Sanatorium as a park and cultural forest in a balanced way that takes landscape values into account. The association will continue to preserve the forests for the use of current and future users of the sanatorium, especially from the perspective of cultural tourism and recreational use by local people (Facebook pages of Parantolan metsät ry, 16 September 2019).

In 2021, the association signed a twenty-five-year agreement to lease the land from Metsähallitus, the area around the so-called Lemmenlampi (or 'Love Pond'), which comprises six hectares. Metsähallitus is a state-owned enterprise, and its general task is to use, care for and protect state-owned land and water areas in Finland. Metsähallitus is expected to operate profitably, and the forests near the sanatorium are included in its commercial forest area. The lease agreement considers the integrity and recreational use of the forest, at least for the present. The area is valuable not only from an environmental standpoint, but also in the eyes of locals because it contains the remains of sanatorium activities. For example, there are still structures in the Lemmenlampi area designed by Alvar Aalto (TYKL/aud/1851) (Figures 3.4 and 3.5).

In the same year, 2021, the association opened a 2-kilometre cultural trail through the forest, called the Sanatorium Path. The trail runs through the association's pine woods and the terrain leased from Metsähallitus around Lemmenlampi. The trail includes signs, information boards and benches, all built by volunteers and with external funding. Another guided trail, the Preitilä Route, was opened in the spring of 2023. This path links the Sanatorium Path with the Paimio Nature Trail via a new connecting path. The city of Paimio oversaw the project, but a group of local volunteers also helped create the route. Volunteers also maintain a tennis court belonging to the association (Facebook pages of Parantolan metsät ry, 31 May 2021 and 22 May 2023). Today, the future of the Paimio Sanatorium's nearby forests seems rather secure. From the perspective of the agencies involved in the heritagization of the site, however, the situation is more complex.

Figure 3.4 The dam at Lemmenlampi, Love Pond in English, in August 2023 (Photo: Maija Mäki 2023).

Figure 3.5 The water pumping station by the side of Lemmenlampi, Love Pond in English: the picture also shows the new information board made by the local association Paimion parantolan metsät ry (Photo: Maija Mäki 2023).

Ambivalent perspectives: The agency of locals in relation to national and global goals

According to sociologist John Urry (1995: 152), often various combinations of and overlaps between global and local processes co-exist in heritagization work. The processes can create possibilities for making local heritage visible and accessible, and create new resources for the economic and social growth of the sites and neighbourhoods. From both a national and a global standpoint, the significance of the Paimio Sanatorium and its forests is unquestionable. Already in 2005, the Finnish National Board of Antiquities prepared a proposal for UNESCO to add the Paimio Sanatorium to the UNESCO World Heritage List (Ehrström et al., 2005). In a new proposal submitted to UNESCO, the Paimio Sanatorium is counted as one of twelve sites designed by Alvar Aalto that deserves an official heritage designation (Wessman, 2019). Aalto's architecture had a strong global impact on modernism, and the Paimio Sanatorium was his and his wife Aino Aalto's first major work. The buildings of the sanatorium are currently protected by the Building Protection Act, which is a law that strongly governs land use in Finland. Aino and Alvar Aalto made a conscious decision to include the forest landscape in the sanatorium's garden composition (Saarikko, 2007: 64). They made the decision not only for therapeutic reasons, but also because the pine forest landscape was of artistic and aesthetic significance for the sanatorium complex.

In the eyes of the museum authorities, the forests of the sanatorium have been an integral part of the sanatorium complex. Museum authorities regarded the cultural landscape of the sanatorium as extending even further than the actual forest zone (Ehrström et al., 2005). The owner of the sanatorium complex, the Hospital District of Southwest Finland, viewed the forests differently and sold them to a private owner as commercial forests. This was possible because forests are not protected by legislation like the built environment in Finland. The locals understood that cutting down forests for commercial use would compromise the success of the UNESCO application (e.g. TYKL/aud/1850; TYKL/aud/1851). Prior examples exist of UNESCO removing sites from the World Heritage List (Heiskanen and Romsi, 2021).

Locals have long been aware that the sanatorium is an internationally significant destination. However, there has been little debate about what this means specifically for the local community. Rodney Harrison (2013: 138–9) has noted that the relationships between local actors and global processes produce creative spaces of friction that might have long-lasting consequences for both local actors and global institutions. Such moments of friction are essential factors for understanding the nature of ambivalent heritage. The heritagization of the site is not only in the hands of authorities, such as museum professionals or bureaucrats, but the public debates generated around specific heritage issues can help define, refine and develop the models and ways of interpreting cultural heritage (Krauss, 2008; Harrison, 2013: 138). The negotiations that take place during heritagization processes are sometimes imperceptible, but at other times are quite visible, which helps highlight the frictions caused by ambivalent heritage.

Critics have argued that heritagization processes, especially when in close contact with tourism and commercialization, have led to the loss of a sense of place at many sites (e.g. Walsh, 1995). Cultural heritage products that have lost a unique identity and connection to local residents can appear false and kitschy. In the case of the Paimio Sanatorium, one may ask what kind of destination it would be without its surrounding forests.

In the case of the forests around the Paimio Sanatorium, the heritagization process is intertwined with the well-known architectural history of the sanatorium, the oral history of the local community, and the needs and wishes of the Paimio Sanatorium Foundation and locals living on the site at present. By purchasing the surrounding forests, locals have concretely taken over a site that they define as important, and consequently, from the perspective of ambivalent heritage and memory activism, the locals have created a stronger sense of agency with respect to the future than before. The new association is able to act as a partner in the development processes in the sanatorium area, and ownership of the land has also given the association more power to influence decisions concerning the area and its future. Experts have observed that the development of cultural tourism is made possible by effective cooperation between a wide range of actors. Although successful cases include authorities and municipalities, they also often require the efforts of local actors working through associations and in cooperation with businesses (Mäki, 2020: 291–3).

Currently, there are several new plans and ideas to develop the sanatorium area as a cultural tourism destination. It is already a popular tourist destination that also offers overnight accommodation. The site includes a restaurant and visitors can also partake in exhibitions, guided tours and events. In 2023, the Paimio Sanatorium Foundation established the Institute for Paimio Sanatorium, which offers, for example, seminars and other events that deal with well-being and sustainability, as well as responsible building and architecture (Paimio Sanatorium web pages, 2023). The foundation is striving to become a globally respected agent, and the main target groups are architects and designers from all over the world. However, it also plans to target locals; in the summer of 2023, the foundation held a summer celebration at the sanatorium aimed specifically at local residents.

When considering the different manifestations and levels of agency, it is important to understand that agency is a socially constructed concept, created in relation to societal structures. From the perspective of the local community, agency can be understood as a small-scale activity that is unobtrusive and connected to people's everyday lives (Honkasalo, 2004). For example, in the case of Paimio Sanatorium, the local residents have devoted countless hours to renovating the walking trails and posting signs and information boards.

> We are enthusiastic and have a lot of ideas. It is a guarantee of success. People will come along; someone just needs to be active. But at the same time, I think people are cautious; they are afraid of failure and [making] mistakes. Today in social media, there will always be someone who is barking that we are doing something in the wrong way.
>
> TYKL/aud/1850

The local community has not been unanimous on the matter. The interviewees talked about the reputation they have gained from participating in protection disputes. Some of the locals disagree with interfering in the affairs of landowners, even though in some cases it would be worthwhile to protect a historical site. In fact, being an active debater on conservation issues can locally earn you the reputation of being 'a troublesome' person (TYKL/aud/1848; TYKL/aud/1851). The community witnessed many internal disputes during the project to save the forests, but afterwards common consensus was reached that it was good to have the forests protected (TYKL/aud/1850).

> 'There was a need for such a foolishly brave person, and on the other hand the particular [type of] person who knew laws and regulations, so that no one could complain about anything later.'
>
> TYKL/aud/1851

In this case, the members of the group had, by chance, the expertise to meet the very needs that arose during the protection process. At the same time, such individuals can be seen as bearers of the vernacular heritage of the local community. Through their volunteer work, they made visible the oral history and local values pertaining to the sanatorium, which otherwise would not have come to light. Silent knowledge, unobtrusive everyday actions and previously uncollected oral history emerge and play a socially significant role in cases where protection processes are not undertaken by the authorities but by the local people. However, the protection process for the Paimio Sanatorium has served both national goals and the internationally acclaimed cultural significance attached to the site. Based on this case, the locally sourced heritagization process has played an important role in other dimensions of defining cultural heritage as well.

Between private and collective: Local perspectives on the heritage of the Paimio Sanatorium

When the Paimio Sanatorium was founded, it was a white-coloured, modern hospital complex that stood out clearly from the surrounding environment. Its appearance in the traditional rural landscape might have caused confusion among the local people at the time. The isolation of the new sanatorium from the rest of Paimio was further emphasized by the barbed-wire fence built around it. Contacts between the locals and those at the sanatorium were limited. However, the sanatorium provided a variety of jobs for the locals during construction and later.

When the closure of the hospital became topical, it raised fears among the city officials of Paimio and other locals. The international and national importance of the site was clear to the locals, but its maintenance and future use as a tourist site were viewed as difficult tasks to manage. For the people who did not work at or have any connections to the hospital, the sanatorium remained quite unfamiliar for a long time, as it was not easy to visit the hospital because of strict privacy regulations. In the past, foreign visitors have been able to admire the sanatorium mainly from the outside. The everyday life of the sanatorium and the memories of the locals were rather hidden and overshadowed by the famous husband and wife architect team, Alvar and Aino Aalto, and their design. The contradictions are evident in the concrete way in which ambivalent heritage can create material manifestations. Also, the debate concerning the privacy of the former patients in the case of the sanatorium's photographic collection, highlights the ambivalence of the site's materiality with respect to questions about intellectual and moral ownership and accessibility.

Cases involving sanatoria, hospitals and similar institutions throughout Finland have given rise to vital discussions about the sensitivity and ethics of research concerning, for example, former patients and their privacy issues (e.g.

Baldwin, 2005; Häkkinen and Salasuo, 2015; Kallinen et al., 2015; Kuuva, 2018; Jäntti, 2022). The various aspects of ethics and issues of sensitivity regarding the dark and difficult history of heritage sites have prompted much discussion in critical heritage studies as well (e.g. Stone, 2006; Roberts and Stone, 2014; Koskinen-Koivisto and Thomas, 2017). The currently strict personal data restrictions and General Data Protection Rules (GDPR), which restrict the public from using material that includes personal data, have influenced how the sanatorium's past can be brought to light today. In 2021, when the sanatorium was first opened to the public, the exhibitions and former patient rooms did not contain any actual information about daily life at the sanatorium. No photographs, memoirs or personal belongings of former patients or staff members were on display. The only photos, letters and other personal items belonged to Aino and Alvar Aalto (Heimo, 2022). Nowadays, the situation is still quite the same for the exhibitions, but guided tours have made use of oral history materials to create new kinds of experiences at the site.

Turku's Lazaret Museum contains a large collection of photographs concerning the Paimio Sanatorium. The photographs are legally owned by the Hospital District of Southwest Finland, but many of them were donated by the local people of Paimio. The photographs fall within the scope of patient privacy and therefore could not be put on public display. The locals had access to a few photographs for information boards on the cultural path, but the patients' faces are blurred. Those photographs are unique, and thus far the cultural path through the pine forest offers visitors the only possibility to learn about daily life at the sanatorium with the help of old photographs. The case reveals an interesting ethical and moral dilemma having to do with the relationship between safeguarding the anonymity of patients and the right to be remembered and share one's experiences with a serious illness. The locals would like to share more information, more memories and more old photographs with visitors, but so far this has not been possible. However, the present oral history project has collected and archived local memories of persons and events, utilizing such memories in many different ways, not only for research purposes but also for public use, such as developing the new guided tours. Many unanswered questions remain, and ethical considerations must be made on a case-by-case basis. Based on the experiences with this case, more research is needed regarding sensitive heritage issues and collecting, interpreting and presenting ethically challenging research materials within the context of vernacular and ambivalent heritage.

Sanatorium without forests?

'The forest is the home of the sanatorium' was the main message that locals wanted to express after successfully purchasing the pine forests surrounding the Paimio Sanatorium (TYKL/aud/1848). The forests around the sanatorium constitute a forgotten vernacular heritage precisely because the safeguarding actions focused so strongly on the built heritage of the site. However, the daily life of the sanatorium extended into the forests, where physical remains of the sanatorium's activities can also still be found. Many sources, including patient interviews and reminiscences, as well as the *Pasuuna* magazines made by the patients themselves, document the significance of the forests for patients at the sanatorium and their experiences in the forests.

Unspoken memories stay with people over time and create a local consciousness connected to particular sites. These hidden memories may include bodily and emotional memories and new information about the site, which only become apparent during heritagization processes. At sites perceived as having difficult or dissonant histories, such as hospitals, documenting and performing oral history are ways of breaking down stigmas and prejudices related to a certain disease (Hakosalo, 2021; Havula, 2021; Jäntti, 2022). In the case of the Paimio Sanatorium, there are hidden stories concerning the role of forests in everyday life at the sanatorium. The staff at the sanatorium exercised power over the patients and denied them many things, but patients in the woods had their own agency. Though locals may have heard stories about the woods in rough detail, the more official public narratives of the sanatorium have dominated the story of the famous architects and internationally significant design and architecture of the site. The protection process for the forests gave new opportunities to discuss aspects of everyday life at the sanatorium. Today, the narratives of daily life at the sanatorium, such as 'lumpustelu' in the forests, are shown in the forests, recorded on the information boards beside the cultural path around the sanatorium.

The heritagization of the sanatorium has shown in a concrete way the importance of acknowledging the agency of local people and their ability and willingness to define the heritage of their own environment, not only through the built environment but also in terms of the natural environment of the site. The process of questioning the ownership of the lands around a nationally and internationally significant site, offers a strong opportunity for empowering locals to influence the future of the site's development. At the same time, the vernacular heritage of the sanatorium has been a fruitful source of protective activism by locals. Collaboratively working with them through the collection and use of oral history materials has strengthened especially their mental ownership of the site. Finally, this feeling of ownership is an important feature of identifying and interpreting the agencies involved in heritagization processes.

References

Archives of the School of History, Culture and Arts Studies (SHCAS), University of Turku, TYKL/aud/1848–1851:

TYKL/aud/1848: 'Teacher' and 'Local Politician'

TYKL/aud/1849: 'Forester' and 'Lawyer'

TYKL/aud/1850: 'Entrepreneur'

TYKL/aud/1851: 'Architect'.

Aalto, A., M. M. Sukkinen, I. Kalkas and S. Savonen (1933), *Varsinais-Suomen tuberkuloosiparantola*, Turku: Preitilä.

Ashton, P. and H. Kean (2008), *People and Their Pasts*, London: Palgrave Macmillan.

Bakalarz-Duverger, M. (2023), 'Communities', in Y. Gutman and J. Wüstenberg (eds), *The Routledge Handbook of Memory Activism*, 91–5, London: Routledge.

Baldwin, C. (2005), 'Narrative, Ethics and People with Severe Mental Illness', *Australian and New Zealand Journal of Psychiatry*, 39 (11–12): 1022–9.

Burke, S. (2018), *Building Resistance: Children, Tuberculosis, and the Toronto Sanatorium*, Montreal and Kingston: McGill-Queen's University Press.

Ehrström, M., S. Jetsonen and T. Lindh (eds) (2005), *Nomination of Paimio hospital for inclusion in the World Heritage List*, National Board of Antiquities, Department of Monuments and Sites, report no 13. https://www.museovirasto.fi/uploads/Arkisto-ja-kokoelmapalvelut/Julkaisut/nomination-of-paimio-hospital.pdf.

Finnish Heritage Agency (2009), 'Museovirasto, Valtakunnallisesti merkittävät rakennetut kulttuuriympäristöt', 22 December 2009. Accessed 10 June 2023. http://www.rky.fi/read/asp/r_kohde_det.aspx?KOHDE_ID=1795.

Finnish Heritage Agency (2021), 'Suomi esittää Alvar Aallon arkkitehtuuria Unescon maailmanperinnön aieluetteloon Museoviraston valmistelun pohjalta', *Museoviraston uutiset*, 1 February. Accessed 19 June 2023. https://www.museovirasto.fi/fi/ajankohtaista/uomi-esittaa-alvar-aallon-arkkitehtuuria-unescon-maailmanperinnon-Aieluetteloon-museoviraston-valmistelun-pohjalta.

Haenn, N. and D. G. Casagrande (2007), 'Citizens, Experts, and Anthropologists: Finding Paths in Environmental Policy', *Human Organization*, 66 (2): 99–102.

Häkkinen, A. and M. Salasuo (2015), *Salattu, hävetty, vaiettu: Miten tutkia piilossa olevia ilmiöitä*, Vastapaino: Tampere.

Hakosalo, H. (2020), 'Tuberculosis Sanatorium Patients' Perceptions of the Healing Power of Nature', in Ruuskanen, E. and H. Hakosalo (eds), *In Pursuit of Healthy Environments: Historical Cases on the Environment-Health Nexus*, 63–83, London: Routledge.

Hakosalo, H. (2021), 'The Ill(s) of the Nation: The Experience of Tuberculosis in Finland from the 1920s to the 1970s', in V. Kivimäki, S. Suodenjoki and T. Vahtikari (eds), *Lived Nation As the History of Experiences and Emotions in Finland, 1800–2000*, 241–66, London: Palgrave Macmillan. https://doi.org/10.1007/978-3-030-69882-9_10.

Harrison, R. (2013), *Heritage. Critical Approaches*, London and New York: Routledge.

Havula, P. (2021), *'Jokanen oli masentunu ku tuli sinne: Ku aatteli, et nyt tää elämä oli tässä.' Muistitietotutkimus Paimion parantolan potilaiden kokemuksista 1940–1960-luvuilta*, Master's thesis, Folkloristics, Turku: University of Turku. https://urn.fi/URN:NBN:fi-fe2021052731861.

Heikinheimo, M. (2016), *Architecture and Technology: Alvar Aalto's Paimio Sanatorium*, Helsinki: Aalto University. https://aaltodoc.aalto.fi/handle/123456789/19607.

Heimo, A. (2022), 'To Display or not to Display? Personal Memories from Paimio Sanatorium', *Ethnologia Fennica*, 49 (1): 91–5. doi: https://doi.org/10.23991/ef.v49i1.121477.

Heiskanen, H. and V. Romsi (2021), 'Unesco poistaa Liverpoolin satama-alueen maailmanperintölistalta – uudisrakentaminen ja 500 miljoonan punnan stadion johtaneet "historiallisen arvon menetykseen"', *Yle Uutiset*, 21 July 2021, https://yle.fi/a/3-12029433, accessed 29 June 2023.

Honkasalo, M.-L. (2004), '"Elämä on ahasta täällä": Otteita maailmasta, joka ei pidä kiinni', in M.-L. Honkasalo, T. Utriainen and A. Leppo (eds), *Arki satuttaa: Kärsimyksiä suomalaisessa arkipäivässä*, 51–81, Tampere: Vastapaino.

Jaakkola, S. (2015), 'Paimion parantola sotasairaalana', in S. Pirhonen (ed.), *Pääteasema: Karjalaiset Paimiossa*, 72–7, Paimio: Paimion Karjalaseura.

Jäntti, S. (2022), 'Muistoja keräämässä ja tutkimassa', in S. Jäntti, K. Heimonen, S. Kuuva, K. Maanmieli and A. Rissanen (eds), *Kokemuksia mielisairaalasta: Muistoihin kaivertuneet tilat*, 32–64, Suomalaisen Kirjallisuuden Seuran Toimituksia 1483, Helsinki: Suomalaisen Kirjallisuuden Seura.

Jäntti, S., K. Heimonen, S. Kuuva, K. Maanmieli and A. Rissanen (2022), 'Mielisairaalamuistot nykykulttuurissa', in S. Jäntti, K. Heimonen, S. Kuuva, K. Maanmieli and A. Rissanen (eds), *Kokemuksia mielisairaalasta: Muistoihin kaivertuneet tilat*, 11–31, Suomalaisen Kirjallisuuden Seuran Toimituksia 1483, Helsinki: Suomalaisen Kirjallisuuden Seura.

Jokela, L. and E. Pihkala (2019), 'Parantolan metsään ei tule taloja – Paimion parantolasta halutaan maailmanperintökohde', *Turun Sanomat*, 25 September 2019, https://www.ts.fi/uutiset/4712715, accessed 28 June 2023.

Kallinen, K., H. Pirskanen and S. Rautio (2015), *Sensitiivinen tutkimuksessa: Menetelmät, kohderyhmät, haasteet ja mahdollisuudet*, Tallinn: United Press Global.

Kelly, S. (2011), 'Stigma and Silence: Oral Histories of Tuberculosis', *Oral History*, 3 (1): 65–76.

Kokko, T. (Forthcoming), *Ei mitään yöpaidassa laahustelua. Potilaiden toimijuus tuberkuloosiparantolassa 1930–1950-luvuilla*, Master's thesis, Ethnology, Turku: University of Turku.

Korjonen-Kuusipuro, K. (2012), *Yhteinen Vuoksi – Ihmisen ja ympäristön kulttuurinen vuorovaikutus Vuoksen jokilaakosssa 1800 – luvulta nykypäiviin*, dissertation, Acta Univ. Oul. B 109, Oulu: University of Oulu.

Koskinen-Koivisto, E. and S. Thomas (2017), 'Lapland's Dark Heritage: Responses to the Legacy of World War II', in H. Silverman, E. Waterton and S. Watson (eds), *Heritage in Action*, 121–33, Cham: Springer. https://doi.org/10.1007/978-3-319-42870-3_9.

Krauss, W. (2008), 'European Landscapes: Heritage, Participation and Local Communities', in B. Graham and P. Howards (eds), *The Ashgate Research Companion to Heritage and Identity*, 425–38, Aldershot and Burlington: Ashgate.

Kuuva, S. (2018), 'Mielisairaalan varjoista: Nostalgia ja melankolia mielisairaala-alueella asuneiden lapsuusmuistoissa', *J@rgonia*, 16 (31): 12–40. http://urn.fi/URN:NBN:fi:jyu-201805302901.

Latour, B. (2005), *Reassembling the Social: An Introduction to Actor–Network–Theory*, Oxford: Oxford University Press.

Mäki, M. (2020), *Polkuja esihistoriasta tulevaisuuksiin – matkailun tulevaisuuksia ja toimijoita arkeologisilla maastokohteilla*, Kansatieteellinen Arkisto 61, Helsinki: Suomen Muinaismuistoyhdistys.

Meurling, B. (2022), 'Från klövrande till rysk diplomati: Om relationer på sanatorier', *Kulturella Perspektiv* 2022, 31: 1–11.

Nieminen, A. (2022), 'Ounasjokiliike', in A. Nieminen (ed.), *Luonnostaan eläväinen: Kittilän historia*, 298–9, Kittilä: Kittilän kunta.

Paimio Sanatorium, web pages (2023), https://paimiosanatorium.com/, accessed 30 June 2023.

Paimio Sanatorium Foundation (2023), https://paimiosanatorium.com/sanatorium/foundation/, accessed 20 June 2023.

'Parantolaperinnettä 9000 sivua' (1972), *Jousi: Tuberkuloosi- ja keuhkovammaisten liiton lehti*, 27 (1): 6–8.

Parantolan metsät ry, Facebook pages (2019, 2021, 2023), https://www.facebook.com/profile.php?id=100064693030572, accessed 29 June 2023.

Pesonen, N. (1980), *Terveyden puolesta – sairautta vastaan: terveyden- ja sairaanhoito Suomessa 1800- ja 1900-luvulla*, Porvoo: WSOY.

Portelli, A. (1997), *The Battle of Valle Giulia: Oral History and the Art of Dialogue*, Madison: University of Wisconsin Press.

Roberts, C. and P. Stone (2014), 'Dark Tourism and Dark Heritage: Emergent Themes, Issues and Consequences', in I. Convery, G. Corsane and P. Davis (eds), *Displaced Heritage: Responses to Disaster, Trauma, and Loss*, 9–18, Suffolk Boydell & Brewer. doi:10.1017/9781782044109.003.

Robertson, I. J. M. (2016), *Heritage from Below*, London and New York: Routledge.

Ruotsala, H. (2002), *Muuttuvat palkiset: elo, työ ja ympäristö Kittilän Kyrön paliskunnassa ja Kuolan Luujärven poronhoitokollektiiveissa vuosina 1930–1995*, Kansatieteellinen Arkisto 49, Helsinki: Suomen Muinaismuistoyhdistys.

Ruotsala, H. (2010), 'From frame to focus: A reflexive journey into cultural environmental research', in J. Lehtonen and S. Tenkanen (eds), *Ethnology in the 21st Century: Transnational Reflections of Past, Present and Future*, 271–86, Scripta Ethnologica Aboensia, Turku: School of Cultural Research, Ethnology.

Ruotsala, H. (2013), *Ounasjoki virtaa vapaana – 30 vuotta Ounasjokilain voimaantulosta*, accessed 2 July 2023. http://serendipitybyanethnologist-helena.blogspot.com/2013/07/ounasjoki-virtaa-vapaana-30-vuotta.html.

Saarikko, J. (2007), *Paimion sairaalan ympäristön historiaselvitys ja kunnostussuunnitelma*, Master's thesis, Landscape Architecture, Espoo: Aalto yliopisto.

Salo, K. (2018), 'Parantolan metsät kaupunginhallituksen hyppysissä', *Kunnallislehti* 19 February 2018, https://www.kuntsari.fi/2018/02/parantolan-metsat-kaupunginhallituksen-hyppysissa/, accessed 29 June 2023.

Savonen, S. (1936), 'Euroopan nykyinen keuhkotautikartta ja tietoja Suomen tähänastisesta keuhkotaudin vastustamistyöstä', *Duodecim* 3/1936: 203–22, https://www.duodecimlehti.fi/lehti/1936/3, accessed 19 June 2023.

Shaw, A. and C. Reeves (2009), *The Children of Craig-y-nos: Life in a Welsh Tuberculosis Sanatorium, 1922–1959*, Milton Keynes: Welcome Trust Centre.

Smith, L. (2007), 'Empty Gestures? Heritage and the Politics of Recognition', in H. Silverman and D. F. Ruggles (eds), *Cultural Heritage and Human Rights*, 159–71, New York: Springer.

Sonck-Rautio, K. (2019), *The Fishers of the Archipelago Sea – Resilience, Sustainability, Knowledge and Agency*, Turun yliopiston julkaisuja – Annales Universitatis Turkuensis, Sarja – Ser. B Osa. – Tom. 493, Turku: Humaniora.

Stone, P. (2006), 'A Dark Tourism Spectrum: towards a typology of death and macabre related tourist sites, attractions and exhibitions', *TOURISM: An Interdisciplinary International Journal*, 54 (2): 145–160.

Suopajärvi, L. (2001), *Vuotos- ja Ounasjoki-kamppailujen kentät ja merkitykset Lapissa*, Acta Electronica Universitatis Lapponiensis 28, Rovaniemi: Lapin yliopistokustannus.

The Petition to acquire Sanatorium forests for the city of Paimio (17 March 2019), https://www.adressit.com/allekirjoitukset/vetoomus_parantolan_metsien_hankkimiseksi_paimion_kaupungin_omistukseen/, accessed 28 June 2023.

Tilastokeskus. Tautikuolleisuus 1936–2010, https://www.stat.fi/til/ksyyt/2010/ksyyt_2010_2011-12-16_kat_003_fi.html, accessed 19 June 2023.

Törrönen, S. (1983), *Varsinais-Suomen tuberkuloosipiiri: Kalevanniemen lastenparantola 1920–1962: Paimion sairaala 1933–1983*, Turku: Varsinais-Suomen tuberkuloosipiirin kuntainliitto.

Tukia, T. (2019), 'Pro parantola-ryhmä vetoaa, että metsät säästyisivät', *Kunnallislehti* 28 Fubruary 2019, https://www.kuntsari.fi/2019/02/pro-parantola-ryhma-vetoaa-etta-metsat-saastyisivat/, accessed 29 June 2023.

Urry, J. (1995), *Consuming Places*, London: Routledge.

Walsh, K. ([1992] 1995), *The Representation of the Past: Museums and Heritage in the Post-Modern World*, London and New York: Routledge.

Wessman, S. (2019), *Raportti aieluettelotyöryhmän työstä*, Helsinki: Museovirasto. https://www.museovirasto.fi/uploads/Kansainvalinen_toiminta/museovirasto_maailmanperintokohteiden_aieluettelo_raportti.pdf.

4

Masculinity, Smokestacks and the Decline of Industry: Visualizing Industrial Heritage in Social Media

Tuuli Matila

Introduction

People are attached to the relics of the industrial past as part of their local identity. Finnish industrial heritage and memories have become especially current in the wake of deindustrialization, as city councils plan to take down old smokestacks and people object to their removal. For instance, in Lahti, a small group of people gathered to object to the removal of an old glass factory smokestack because of its symbolic value (Bavard, 2023). News of demolitions are often met with heated arguments in the comment sections of online stories from people both against and for the destruction of these heritage landscapes (e.g. Rimpiläinen, 2019). Such emotive discourses illustrate the ambivalent nature of industrial heritage.

There are many different strategies to adjusting to the changing landscape because not everything can be saved, obviously. I am involved in the project 'Smokestack Memories', in which my colleagues collected memories from the public related to Finnish industrial remnants. As part of this project, people could add images of smokestacks and narrate their own personal memories on an online map made specifically for the purpose of collection memories and experiences about these industrial structures (www.muistotkartalla.fi). There were quite a few memories that people attached to smokestacks, and they had photographed many of the old factories around Finland. This led me to think about the visual ways to approach industrial heritage and what the smokestack as a visible landscape feature connotates.

The material world around us has changed, and a lot of culture is produced online, especially in social media. Memorialization occurs online in addition to the actual heritage sites, and photography has become an important way of structuring the world around us (e.g. Giaccardi, 2012; Matila, 2021; 2022). For instance, urban exploration is one manifestation of this newly found appreciation and affection towards industrial structures and making sense of their liminal status in the world, and photography is a crucial part of the activity (Martínez and Laviolette, 2016; Matila et al., manuscript). The meanings and uses of old industrial spaces go through substantial change as they no longer fulfil the function that they were originally built for.

Photographs of industrial heritage abound on Instagram, and industrial smokestacks have also been paid special attention. There is indeed a whole category of images in social media, more than 700 images of industrial smokestacks appear with the search #piippu (Finnish for pipe, chimney or smokestack). The images depict smokestacks from various angles, as part of the landscape or individually, in various different atmospheres, ranging from the gloomy autumn skies to the sunny spring surroundings.

My initial hope was that I would see some very distinctly gendered ways that industrial sites were approached, so I started examining the differences in the ways that women and men portray industrial structures in both social media and in the map site that the 'Smokestack Memories' project created. The experiences and discourses around (industrial) heritage tend to be distinctly gendered (e.g. Reading, 2015; Grahn and Wilson, 2018; Taksa, 2020). Obviously, not everyone's gender is readily apparent, so there is a slight uncertainty in my data, although I did categorize the most uncertain subjects as unidentified, 'U/I'. It is important to acknowledge that gender is a fluid category and is full of individual variation. There are, however, various cultural ideas of gender that may still be very rigid (Parente-Čapková et al., 2015; Siltala, 1994). The more I dug into the depths of the pictures and the more categories I came up with, I noticed that I could not say anything conclusive about gendered ways of seeing the industrial past and the smokestack. I looked at framing, atmosphere, distance, lighting, positioning, etc. At first, I was disappointed, but then I realized that perhaps this ambivalence is my result rather than an unfortunate failure of the data to reveal gendered ways of seeing the industrial past. This chapter focuses specifically on the smokestack as a monumental trace of the industrial past, in which various cultural emotions and understandings and gendered history crystallize. This, I argue, results in an ambivalence that is notable in the visual representation of smokestacks.

I research the world visually, but with an emphasis on experience and emotion alongside visual representation. Emotion has been given far less attention in visual research, as photography tends to be approached as historical representations devoid of any experiential character (e.g. Matila, 2022). Once you start to examine images from both representational and emotive perspectives, you realize that they are sometimes quite contradictory and difficult to interpret, as I will point out further on in this chapter. My arguments surround questions of (1) how the industrial environment, and factories are seen photographically, (2) how the industrial landscape and the smokestacks are experiences as things, and (3) how masculinity is intertwined in industrial architecture. In other words, I examine the smokestacks as part of our physical but also mental landscapes. In addition to the content analysis of the smokestack images, I performed qualitative visual analysis, specifically cultural semiotics, which allows me to introduce individual photos with particularly interesting set-ups. Memoirs and the associated images that the public has posted on the project's map application will complement the data.

Finnish industrial history and nationalism

Finns are very proud of their former industry and the quality products that were made in their country, from textiles to ceramics to Fiskars scissors and other items. Finnish forestry production was key to building the welfare state, referred to with the idiom 'green gold', which you often hear in Finnish public discourse. Building the welfare state was a common project that lifted Finland among the wealthy Western countries. Historically in Finland, depicting industries in picture books has been a way of illustrating economic wealth (Häyrynen, 2005: 72).

Unlike most European countries, Finland did not go through industrialization by the nineteenth century, but rather the late nineteenth century, and it was not really until after the Second World War that Finnish society went from an agricultural country to a more modern industrial state with its own working class (Melin, 2020). The shift from industrial production to service production also took place very rapidly from the 1950s and 1960s to the 2010s (Hannikainen and Eloranta, 2019). Because industrialization happened rather late, it is still very close and perhaps affective, also on a national level because industrialization is deeply intertwined with Finnish nation-building.

Finland is also unique in that deindustrialization has not been very dramatic. Industrial production is still important, but the levels of production have dropped significantly. From its heyday in 1974 when industry employed some 552,000 workers, it now employs some 320,000 workers (Hoffman, 2019: 152).

Metal manufacturing, chemical and electronics industries have been sustained, whereas forestry has significantly decreased, and the textile industry is all but gone (Hoffman, 2019: 157–60). The oldest industries are what really 'built the country', so their importance for the 'national psyche' are perhaps greater than those that have been sustained.

Photobooks are especially fruitful when examining nationalist representations of industry. The (i.e. 1942) photobook *Das ist Suomi* (This is Finland), which was directed to a German audience, illustrates this point well. The German audience was important for Finns, and they cared about how they would be perceived in Germany as Finns and the Nazis were waging war together, against the Soviets in the Eastern front in the Second World War. The Nazis had categorized Finns as a Mongolian race, so the Finns were quite eager to counter these perceptions (Kemiläinen, 1998; Häyrynen, 2005; Koponen et al., 2018). Three pages in this photobook depict Finnish factories and list industries ranging from textiles and ceramics to forestry production. What is important here is the way these factories are represented, as one caption in the book states: 'Generous Finnish industrial beauty. The lines of the factory – a straight profile, symbolizes the assurance and determination of the national work performance. Naturality, beauty, and expediency are the mottos of Finnish work' (Suova 1942). The factory presented is the Sunila forestry product factory in Southern Finland, and the emphasis seems to be on the symbolic and national importance of straight lines. As Hilary Orange (2016: 13) notes (quoting High and Lewis, 2007: 2; Holzberg and Giovanni, 1981: 318, 322), industrialization is accompanied by various cultural signs and symbols, and understandings of appropriate national values and attitudes – it is far more than just a livelihood.

As Tim Ingold (2007: 4) argues, straightness in Western culture has symbolic affinities with moral thought and rationality. The symbolism of the straight line represents the Finnish Lutheran work ethic and Finnish character as straightforward and straight-backed. The ideal Finnish character is someone who takes pride in their hard work and does it without complaining. The myth of the straight-backed and always honest Finn is a commonly reproduced myth in the media. For instance, a Finnish tabloid printed a news story that stated that an Italian company owner was shocked by Finnish honesty when a cleaner had lifted a 5-euro bill from under his desk and left it there for the boss to find (Numminen, 2020). While honesty is of course a desirable quality in a person or nation, such national myths can even become blind spots for nations. I have suggested elsewhere that the myth of the always honest and fair Finn has in part enabled the slow dismantling of the structures of the welfare state – cuts have taken place ever since the 1990s, and yet Finns fail to notice the declining state of social policies, such as public healthcare (Matila, 2022; also Julkunen, 2001: 258–9).

Factories are more than just buildings that produce products, jobs and welfare. They are deeply symbolic and are attached to the national consciousness in ways that are not always apparent. Smokestacks are the most obvious reminders of this national attitude, and this straight standing structure is more than just a vehicle for smoke to travel towards the skies. It can be seen as a representation of Finnish work ethic, national character, and as I will argue later, masculinity. This is perhaps another reason why taking down smokestacks from the landscape can be so emotional for people. Obviously, people get attached to smokestacks for other reasons as well, but it has to be noted that deindustrialization has in some ways taken away Finns' ability to fulfil their roles as good citizens through hard work, which can result in collective confusion and even shame; losing some of the traditional industrial jobs and getting laid-off may influence people's self-esteem. The Finnish collective emotional tendency leans towards shame, and Finns are particularly sensitive to such feelings, which explains why such occurrences are deeply felt (e.g. Laine, 2006; Matila, 2022).

Industry and the environment: A world for both sexes

Factories are a somewhat constant part of Finnish physical surroundings, especially in cities. However, when they first appeared in both rural and urban areas, it was important to justify these huge complexes that instantly changed landscapes. Another picture caption in the 1942 photobook, *Das ist Suomi* states: 'Most of the Finnish wood processing factories are hidden amongst nature and the workers are guaranteed a healthy way of life. The lights of the factory reflect on the surface of the lake and enliven nature rather than damage it' (Suova 1942).

Besides an indicator of moral uprightness, the straight line is also a symbol of the victory over nature (Ingold, 2007: 152). Historically, however, it seems that Finns have represented industry as something that is in harmony with nature. This obviously has some bearing on how these factories are represented in social media, as historical visual traditions often unconsciously influence our understanding of our culture and surroundings (e.g. Häyrynen, 2005: 62–4, 188). There were a notable number of images that depict factories amidst forests, flowers, lakes and sunny skies, and, in general, very picturesque scenes, such as those of Finnish national romantic depictions in various photobooks.

Table 4.1 Image categories according to the assumed gender of the photographer

#piippu (#smokestack)	Amount	Percentage
Dark M	32	32
Dark F	21	21
Dark U/I	1	1
Bright M	18	18
Bright F	34	34
Bright U/I	1	1
Barriers M	39	39
Barriers F	39	39
Distant M	20	20
Distant F	21	21
Close M	27	27
Close F	31	31
Close U/I	2	2
Neutral	11	11
All	100	100

The financial benefits of factories are all but gone. Also, we cannot ignore the fact that these factories are one cause for climate change, and we now feel maybe an ominous feeling when looking at these smokestacks. In the imagery in social media, this is evidenced by the slightly more pronounced emphasis on dark moods and dramatic skies around these factories, or at least gloomy and cloudy atmospheres (Table 4.1). On the other hand, it seems that this is somewhat divided among the sexes; men tend to favour dark and dramatic skies, whereas for women brightness and the sun is more prominent. Since climate denialism, for instance, has been attached to the masculine identity and emotional condition, it is more unlikely that the dark moods are related to a sense of threat (Anshelm and Hultman, 2014). The darkness of atmosphere could result from each of the sexes claiming the factories into their own experience worlds; men emphasize gloomy skies, making the factory a masculine world, whereas women take these places as their own and attribute sunny atmospheres and 'feminine emotions' (see Jung and Griber, 2019 on colour associations according to sexes). Dirt and danger, the signifiers of the masculine industrial world (Wajcman, 1991; Taksa, 2020), are translated into drama and darkness visually (Figure 4.1). Perhaps there is an ambivalence towards the industrial past and a conflict with

Figure 4.1 A factory at night. The image mirrors some of the visual tendencies of men when photographing smokestacks (Photo: Tuuli Matila, 2022).

what it gave people and what it took, and that is visible in the way that people represent factories. Also, both sexes feel like the industrial past is 'theirs'.

As mentioned, my colleagues collected memories related to smokestacks in an online map, where the public could also attach photographs of the factories that were important to them. I analysed these images with the same intent of seeing how the structures were approached visually. I could see that factories that were still functioning were more often portrayed as part of the surrounding landscape, i.e. shot from a distance. It could be because people cannot get near these places, but it also might suggest that they still see them as part of their mindscape and place-bound identity, more so than the ones that do not function anymore. Indeed, many researchers have pointed out that our identity formation often involves places of importance, such as a place of birth (e.g. Taylor, 2010; De Nardi, 2020). This suggests that there is a real emotional impact when factories are closed, which goes beyond just losing jobs. Factories are part of what attached people to places; they can guide them home and be important sensory markers in that way (e.g. Mah, 2009).

The use of brick implies permanence (Leone and Hurry, 1998), since in the past wooden constructs would have burnt easily in city conflagrations. When the industrial buildings were constructed, much of Finland was still made up of timber, and these vast buildings would have had a major impact on people's experience of their landscape. There is a sad irony that the structures are still standing with no operations inside. Factories are like miniature towns, and there is a sense of community that is often lost when these mass employers move production away from the country of origin, and I suggest that this may have some bearing on how the old factories are approached visually. Almost 40 per cent of the pictures have been photographed from behind 'barriers', which can indicate mental distance to the photographed target.

Existentialist and religious connotations

There are softer values at work when considering the importance of the smokestacks, such as how they guide our perceptions and act as anchors of our place-bound identities. It may be a comforting aspect of the landscape, an aide with which to navigate home or a reminder of childhood landscapes (see the next section). However, I hypothesize that these structures would have reminded the workers of self-control, since round, tall structures are built to be seen and 'see' everywhere (Leone and Hurry, 1998). There was one interesting image in social media, where a woman had photographed a smokestack with two red lights that appeared to her like eyes that were watching her as she stated in the caption. She attributes life but also a surveillance experience to the smokestack. Industry was about top-down control (Wurst and Mrozowski, 2016), and the height of the smokestack is therefore relevant in this regard. The industrial world is long gone, but perhaps the sense of the patriarchal social control to some degree remains, and the smokestack reminds us of the 'natural' world order.

Lines are about travelling (Ingold, 2007: 75–84). The smokestacks guide our perceptions towards the sky and because of this, there is perhaps also a religious connotation to the smokestack. In the past the industrial complexes were built by the employer and a church was also commissioned (Alapuro, 1977). Industrial owners were, to a degree, in charge of the religious education of their workers. Also, schools for the workers' children were established where religion was taught (Kortekangas, 1965). All these actions would have diluted the power of the workers' organizations and the associated Marxist socialism that saw the church as inseparable from the capitalist economy that they were fighting against – many demanded that the church be separate from the state and regarded rather as a private matter (Murtorinne, 1995: 32–3). The church and industrial owners had mutual interests in maintaining social order among the workers.

To add to the religious connotations of the smokestack, think of how the Vatican utilizes the smokestack in its centuries-long tradition – Catholics wait nervously for the white smoke to be emitted from the Sistine Chapel's wood-burning stove when the College of Cardinals elects a new pope. Sara Ahmed (2006) has written about queer phenomenology and queer ways of seeing. She argues that there is a certain violence to how straight lines force us to look. According to Ahmed (2006: 14; quoting William James, 1890: 149), 'space in itself is sensational: it is a matter of how things make their impression as being here or there, on this side or that side of a dividing line, or as being left or right, near or far'. The straight line gives us direction, and by pointing towards the sky, symbolically the smokestack reminds us of the divine roots of the patriarchal order (see also next section).

Also, in an experiential way, the smokestacks remind us of our smallness in the world, not dissimilar to the way churches and cathedrals do. There is a whole category of images where the smokestack has been shot from below towards the sky (Figure 4.2). Smallness has been described in a spiritual light, as a self-transcendent experience that helps one overcome self-related concerns and focus on others (Piff et al., 2015; Preston and Shin, 2017). By standing next to the smokestack, people may get feelings of awe and grandeur that affect how they feel about their everyday concerns and perhaps even makes them feel like they are part of larger forces, a greater whole (Piff et al., 2015). Since religion has less meaning in peoples' lives today, they turn to other things around them for spiritual feelings. Urban exploration in abandoned factories is one example of this modern spirituality, and I argue that the smokestack can perform the same function (Matila et al., manuscript). The massive scale of industrial structures is relevant in this regard. A personal connection with heritage is important in ways that may not always be readily apparent.

Monumental masculinity

> *The place feels castrated when they take down the smokestacks. Fuck, it's like the masculine world is also castrated because you cannot say or do anything anymore.*
>
> Comment on www.muistotkartalla.fi page

Figure 4.2 A smokestack from below. The image mirrors the way smokestacks are often photographed in social media (Photo: Tuuli Matila, 2023).

This is one of the memoirs collected by the 'Smokestack Memories' team, where a man reminisced the removal of a smokestack of an old brick factory. He stated that the smokestack had been a part of his childhood landscape and could be seen from a distance above the fields, and often guided the man home. But the man also sees taking down smokestacks as a castration of the landscape, and feels especially nostalgic about the past when it was still 'ok to be a man'. I will now discuss the symbolism of the smokestack from a gendered angle. I borrow here from geography scholarship on verticality, most notably around the skyscraper and its symbolism.

Politics and masculinity are deeply intertwined with the building of skyscrapers. There is an ongoing competition on which nation builds the tallest structure. Currently, the biggest is of course the Burj Khalifa in Dubai, measuring some 828 metres. Psychiatrist Lawrence Blum (2012: 26) describes in the *Psychiatric Times* journal how men develop fascinations with 'cigars, pens, cars, trains, baseball bats, knives, guns and sausages, but not usually with frisbees, soup, pillows or suitcases' (although as frisbee golf is a very popular sport nowadays, so maybe this is not true anymore). Men will always be concerned with size and psychologists would say that this is a result of childhood. The child sees their mother have a relationship with their father that the child is excluded from but desperately wants to reach. So, the child feels that they are trapped in their own smallness and is therefore convinced that bigger is better (Blum, 2012). I do not think I have to explain in depth all the manifestations of this attitude, but needless to say that from popular culture, pornography, etc. it is obvious that men are often concerned with size.

Feminist critics are therefore correct in that the skyscrapers that in twentieth-century urban America climbed towards the sky are symbols of modern masculinity that indeed 'reeked of an extreme, phallic-centred patriarchy' (Graham, 2016: 154). The twentieth-century urban skyscraper architectural commentator Leslie Kanes Weisman (2000: 1–5; cited in Graham, 2016: 154) writes, is 'a pinnacle of patriarchal symbology, and rooted in the masculine mystique of the big, the erect, the forceful—the full balloon of the inflated masculine ego'. I do not think it is a coincidence that Näsinneula, the tallest building in Finland, was built in Tampere, a city known for its industry and many red-brick smokestacks. A picture book from 1973 describes the industrial city of Tampere as 'big and powerful', aside images with smokestacks (Talvi et al., 1972: 93). Tampere is also described as strong, 'the city with broad shoulders', in another picture book from 2009 (Rautio and Kohvakka-Viinanen, 2009: 4). The former industrial town then, is, in a way, the most masculine Finnish city. Näsinneula was built in 1971, which coincides with changes taking place at the Finlayson textile factory, which is the oldest textile factory in town. Some of the more paternalistic aspects of industry were slowly dismantled all through the 1970s and early 1980s, such as the building of workers' housing. Also, the textile and leather industries were faced with more difficult times as global competition increased, so the city was in dire need of more modern, masculine symbols (Koivuniemi et al., 2022: 306–19, 332–5). Näsinneula has been the tallest building in Finland until the very near future, when some architectural agencies are planning to build taller buildings in Kirkkonummi and Helsinki. Subsequently, in a local newspaper it was suggested that Näsinneula's antenna should be lengthened by 18 metres, so Tampere could keep the status of having the tallest building in Finland (Nurmi, 2022). The architects who have planned these tall buildings are all men. Men have always enjoyed industrial technologies and designs that hint at physical capabilities and prowess (Cockburn, 1986: 93, 97–8; Taksa, 2020). Therefore, the industrial smokestack was an apt symbol of the masculine world.

I will briefly venture to examine the Washington monument, which I see as a particularly masculine monument, to argue that not only is the smokestack a sign of masculinity, but it is also a monument of masculine rationality. Firstly, when the construction of the Washington monument ceased for a while, Mark Twain aptly compared the stump of the Washington monument to 'a factory chimney with the top broken off' (Meyer, 2001: 155). There is hence an obvious visual link with the monument and the smokestack. Secondly, there is something very interesting about the symbolics of the monument. In another monument built to honour George Washington, by Horatio Greenough, Washington was portrayed without a shirt, and apparently this caused some commotion among the public since he could not be imagined 'naked' (Meyer, 2001). But, there is some obvious phallos symbology both in the monument and in his portrait in the one-dollar bill – the outline that has been described as 'mushroom-shaped' (Meyer 2001: 132). Washington has a semi-deified status, so imagining him portrayed in the nude, let alone with the actual penis exposed is impossible and would perhaps be too profane. As a result, the suggestions of his masculinity are indirect.

Science, geometry and measurements became traits eagerly applied to urban design in seventeenth-century England (Peckham, 2009). Straight lines then, can be considered as symbols of rationality. Rationality has indeed historically been considered as a predominantly male trait, whereas women have been perceived as irrational and emotional (Ahmed, 2006). Ruling men would be considered rational, and their symbols, like the Washington monument, needed to match those ideals. There are symbolics to this design that influence the way we perceive them and experience them as part of our landscape.

These symbolics regarding the smokestack recently manifested in an interesting way visually in Finland. A Finnish artist Kaj Stenvall, known for his duck-themed paintings, recently started painting more politically motivated artworks. His paintings of Putin have become especially popular and have been noted all around the world (Körkkö, 2023). His latest works include a painting of former Finnish prime minister Sanna Marin, a 34-year-old woman and the youngest ever prime minister to step into the office. Marin is a representative of the Social Democratic Party, a traditionally left-leaning political party. Stenvall painted Marin after the announcement of her stepping down as party leader in April 2023. Interestingly, Stenvall has framed Marin between two red-bricked smokestacks in Tampere, the city where she is from. In the painting, she sits with a subtle smile in a way that is clearly referencing the Mona Lisa. Marin is a young woman who was very highly scrutinized during her duties as prime minister, often on a personal level, which aroused discussions about misogyny in both the Finnish and international press (e.g. Pullinen, 2023). Stenvall's painting can be seen as a critique of the treatment that the young female prime minister experienced, and it is interesting that the critique was delivered with these symbols of the 'old' patriarchal world. Also, the Mona Lisa is considered a symbol of happiness, i.e. the appropriate emotional state for a woman given that women, at least in Finland, are not supposed to show signs of aggression as it is considered inappropriate, and women should instead be passive and submissive (Lagerspetz, 1998: 244–70, 256–62). However, aggression is what motivates human action, and indeed in politics it is necessary if one wants to further their own cause. The left in politics is associated with the feminine, as is the left side of the brain, whereas the right side represents rationality and a more competitive and tough capitalistic attitude (Ahmed, 2006: 13). Taking down smokestacks in the landscape thus becomes more than just removing the symbols of industry.

Think of, for instance, the images of the World Trade Center terrorist attack and subsequent collapse of the Twin Towers, and the images of the so-called jumpers that basically became taboo in the US and were banned from public display. In addition to them obviously illustrating tragic deaths, these images were, according to Jackie Kirouac-Fram (2011), upsetting to Americans because they were incompatible with the heroic narratives of 9/11. The phallos symbolism affords another type of explanation. In European culture, like in Ancient Greece, for instance, the phallos was more than anything a protective symbol (Monick, 1987). Julie Drew (2004) argued that 9/11 pushed America towards a more gendered view of national identity, and right after the tragedy, in the tabloids and newspapers it was implied that America had been far too feminine and had to become more masculine to be safer. 9/11 then can be understood as a failure of masculinity, and arouses all kinds of irrational responses and the following emphasis of hyper-masculinity. Universally the penis symbolizes masculinity and 9/11 was a direct blow aimed at the crotch.

Now through globalization we are again finding ourselves in a somewhat socially and economically unsecure time. We are witnessing the growth of movements that stress hyper-masculinity, such as the far-right movement, which has gained new traction since right-wing populist parties are getting a footing in politics all over the democratized world. I argue that the removal of old industrial structures and the smokestacks must also be understood as a masculinity crisis. This obviously is not revolutionary, since the financial crises of the early 2000s has already been described as such (e.g. Sauer, 2014). As Birgit Sauer (2014) aptly points out, the crisis did not bring down the masculinity associated with neoliberalism, but rather took new forms and adapted. Such a view has not been (to my knowledge) considered in heritage studies and what sort of lessons we might get from that. In heritage work it is crucial to understand people's emotions so they might be met with more understanding when these structures are taken down for instance, and come up with new ways to attach masculinity, pride and a sense of self-worth to heritage.

It would be easy to expect that women would have a more negative view towards the smokestack and what it represents. But in the Finnish case it is not so obvious, and again I turn to psychology and history for answers. Historians of emotion have pointed out that Finnish women have had to assume rather masculine roles throughout history (Näre, 2004). The Second World War forced women to take over both the men's work and the 'more traditional' women's work, such as child-rearing, as the men were fighting in the frontlines. Women developed into self-sufficient survivors and perhaps had to neglect softer emotions and care. Industry provided women with the first opportunities to earn a living on their own, so it weakened the patriarchal control over them, and marriage was no longer a necessity of life. So, industrial patriarchalism was not just a weapon aimed towards women, although it was not completely equal, since women did not get paid as much as men, for instance. But by assuming a more masculine role in society, women could be 'freer'. Therefore, it does not seem that women portray smokestacks in social media as markedly more distant, repellent or dangerous compared to men. As mentioned previously, I examined whether the factories were photographed from behind natural 'barriers', such as hills, fences or bodies of water, which can indicate an occluded or distant viewing position (Table 4.1; Lister and Wells, 2001). Men and women had the exact same percentage of photographs shot from behind such obstacles and no gender difference can be discerned.

The smokestack is an ambivalent structure also because it is at the same time a sign of progress and failure. A few people in social media have photographed the smokestack next to the moon. This can also symbolically hint at how we as a society have reached the moon, largely because our success rests on the shoulders of previous generations as Marx and Engels have pointed out (Marx and Engels, 1975: 170; also Ahmed, 2006). This is well illustrated by one specific smokestack in Tampere that appears in the social media imagery often. The smokestack in question features an artwork (made by artist Mario Merz and revealed in 1994) that depicts the Fibonacci sequence from 1 to 55 with luminous, red tube-lights (Figure 4.3). The Fibonacci sequence can be seen as symbolic of growth, usually

Figure 4.3 The Fibonacci sequence in the smokestack of Turku Energy company (Photo: Tuuli Matila, 2023).

spiritual and divine, because when presented geometrically it forms a spiral (see Herva et al., this volume), but maybe in this case also economic and social growth through mathematical (and masculine) rationality.

And yet, these now vulnerable structures stand alone and as heritage, and are somewhat isolated from our everyday experience. As Tim Ingold (2007: 167) states, 'if the straight line was an icon of modernity, then the fragmented line seems to be emerging as an equally powerful icon of postmodernity'. Following the demolition of symbols of masculinity, we may fear the subsequent loss of rationality, and that can make many people feel uneasy.

Discussion

The points discussed above, I argue, result in the ambivalence in my social media data. There are a lot of complex symbolics and emotions at work with the smokestack. Images transfer smokestacks to a new time and life in the digital world. It is a form of resistance towards the vanishing of this ambivalent world and the past.

Smokestack images are expressions of the meanings and experiences that revolve around these industrial structures. They are makers of identity as both Finns, and as men and women. Social media enables people to participate in the maintenance of heritage and can structure their identities too. It is an unstable time people are living in so perhaps this is the reason for the abundance of photographs of industrial heritage in social media. Economic and social instability causes nostalgic attitudes towards the past (also Kivimäki, 2012).

Neil Silberman (2007: 82) has argued that 'heritage is an ever-changing array of objects and symbols, a complex mosaic of artefacts, images, monuments, and customs that demand our attention and demand that we give some meaning to them'. I argue that sometimes these meanings are already given and yet they avoid our consciousness. They are often the result of complex symbolics and emotions. Therefore, it is important to do visual research on heritage that makes clear some of the things that may be directly in front of our eyes, but we fail to see. As Heidegger ([1927] 1985: 69–75; also Ahmed, 2006) points out, we notice the shape of something when it no longer serves its functional purpose, and it turns into something of a failure; it is no longer an extension of the body. The industrial smokestack has in many cases also ceased to do its original function, and it is now that the ties to the masculine identity become more apparent.

Another factor in the abundance of industrial photographs is the very nature of the old industrial landscape. As Ralf Richter (2017) argues, industrial structures, even the old ones, are vast, and these legacies impose our perception in the landscape. They, and especially the smokestack, still very much command our views as these monuments of the industrial past. Forcefully, the smokestack guides our perception towards it and towards the sky where the functioning factories carry their smoke in a realm beyond our reach.

Smokestacks elicit contradictory emotions that are reflected in the heated debates around their removal or preservation. This chapter aimed at making the smokestack visible; to make it a focus point means to draw attention to the various symbolics that are at work. These pictures of industrial smokestacks in social media illustrate the emotional connection to the past but not a past without conflicts.

Conclusion

Smokestacks are concrete but also very visual things that have gone under-theorized for a long time. They are a constant and yet easily ignored aspect of our everyday worlds. They have often been transformed into heritage, and hence their meaning has changed drastically as they no longer serve as facets of industry. They are very much alive though, in the minds of the people who have inherited them from the past. Given that they have been transformed into heritage, it is the heritage professionals' responsibility to be aware of the symbolic and emotional underpinnings of old industrial structures and to strive to understand and work with these feelings together with the public. The public is of course diverse. But, there are some gendered legacies that are inherited from the past and continue to influence perceptions in the present. Visual research can help dissect some of the associations that are tied to these structures. As signs of masculinity, progress and failure, national identity and local identity, they continue to arouse emotive responses.

References

Ahmed, S. (2006), *Queer Phenomenology: Orientations, Objects, Others,* Durham: Duke University Press.

Alapuro, R. (1977), 'Uskonto ja poliittinen mobilisoituminen maaseudulla', in M. Kuusi, R. Alapuro, and M. Klinge (Eds), *Maailmankuvan muutos tutkimuskohteena: Näkökulmia teollistumisajan Suomeen,* Helsinki: Otava.

Anshelm, J. and Hultman, M. (2014), 'A green fatwā? Climate change as a threat to the masculinity of industrial modernity', *Norma: International Journal for Masculinity Studies*, 9 (2): 84–96. doi:10.1080/18902138.2014.908627.

Bavard, C. (2023), 'Lahdessa järjestettiin mielenilmaus monille rakkaan savupiipun puolesta –"Rakennuksilla on myös symbolista arvoa"', *Etelä-Suomen Sanomat*, 7 February. Available at https://www.ess.fi/paikalliset/5710467.

Blum, L. D. (2012), 'Penis size: A consideration of fantasy and reality', *Psychiatric Times*, 29 (12): 26.

Cockburn, C. (1986), 'The material of male power', in E. Review (ed.), *Waged Work: A Reader*, 93–113, London: Virago.

De Nardi, S. (2020), *Visualising Place, Memory and the Imagined*, London: Routledge. https://doi.org/10.4324/9781315167879.

Drew, J. (2004), 'Identity Crisis: Gender, Public Discourse, and 9/11', *Women and Language*, 27 (2): 71–7.

Giaccardi, E. (2012), *Heritage and Social Media: Understanding Heritage in a Participatory Culture*, London: Routledge.

Graham, S. (2016), *Vertical: The City from Satellites to Bunkers*, London and New York: Verso.

Grahn, W. and R. J. Wilson (2018), *Gender and Heritage: Performance, Place and Politics*, London and New York: Routledge.

Hannikainen, M. and J. Eloranta (2019), 'Palveluiden ja tulonsiirtojen yhteiskunta', in J. Laine, S. Fellman, M. Hannikainen, J. Ojala and M. Autio (eds), *Vaurastumisen vuodet: Suomen taloushistoria teollistumisen jälkeen*, 19–38, Helsinki: Gaudeamus.

Häyrynen, M. (2005), *Kuvitettu maa: Suomen kansallisten maisemakuvastojen rakentuminen*, Helsinki: Suomalaisen Kirjallisuuden Seura.

Heidegger, M. ([1927] 1985), *Being and Time*, Repr., translated by J. Macquarrie and E. Robinson, Oxford: Blackwell.

High, S. and D. W. Lewis (2007), *Corporate Wasteland: The Landscape and Memory of Deindustrialization*, Ithaca, NY: Cornell University Press.

Hoffman, K. (2019), 'Suomen teollisuuden yleiskehitys', in J. Laine, S. Fellman, M. Hannikainen, J. Ojala and M. Autio (eds), *Vaurastumisen vuodet: Suomen taloushistoria teollistumisen jälkeen*, 150–75, Helsinki: Gaudeamus.

Holzberg, C. S. and M. J. Giovannini (1981), 'Anthropology and Industry: Reappraisal and New Directions', *Annual Review of Anthropology*, 10: 317–60.

Ingold, T. (2007), *Lines: A Brief History*, London: Routledge.

James, W. (1890), *The Principles of Psychology*, Vol. 2, London: Macmillan.

Julkunen, R. (2001), *Suunnanmuutos: 1990-luvun sosiaalipoliittinen reformi Suomessa*, Tampere: Vastapaino.

Jung, I. and Y. Griber (2019), 'Colour associations for the words feminine and masculine in nine different countries', *International Color Association (AIC) Conference 2019, Color and Landscape, Midterm meeting: October 2019*, Buenos Aires, Argentina.

Kemiläinen, A. (1998), *Finns in the Shadow of the 'Aryans': Race Theories and Racism*, Helsinki: Finnish Historical Society.

Kirouac-Fram, J. (2011), '"The Most Disturbing Aspects": Apprehending Public Reaction to Photographs of the 9/11 Jumpers', *Radical History Review*, 2011 (111): 131–7. doi:10.1215/01636545-1268749.

Kivimäki, V. (2012), 'Between Defeat and Victory: Finnish memory culture of the Second World War', *Scandinavian Journal of History*, 37 (4): 482–504. doi: 10.1080/03468755.2012.680178.

Koivuniemi, J., J. Peltola, H. Pietarinen, M. Lähteenmäki, A. Lehto and M. Lind (2022), *Finlayson: II (1918–2020), Puuvillan jättiläinen*, Tampere: Vapriikki.

Koponen, T., O. Seitsonen and E. Koskinen-Koivisto (2018), '"Das ist Suomi"', *Ennen ja nyt – historian tietosanomat*, 4 (2018). Available at https://journal.fi/ennenjanyt/article/view/108901/63893.

Körkkö, H. (2023), 'Kaj Stenvallin Putin-maalaukset ovat olleet kansainvälinen hitti jo vuoden – juttu näyttää 15 satiirista kuvaa sodasta', *Helsingin Sanomat*, 20 February 2023. Available at https://www.hs.fi/kulttuuri/art-2000009397828.html.

Kortekangas, P. (1965), *Kirkko ja uskonnollinen elämä teollistuvassa yhteiskunnassa: Tutkimus Tampereesta 1855–1905*, Helsinki: Kortekangas, Paavo.

Lagerspetz, K. (1998), *Naisten aggressio*, Helsinki: Tammi.

Laine, T. (2006), 'Shame on us: Shame, national identity and the Finnish doping scandal', *International Journal of the History of Sport*, 23 (1): 67–81, doi:10.1080/09523360500386450.

Leone, M. and S. D. Hurry (1998), 'Seeing: The Power of Town Planning in the Chesapeake', *Historical Archaeology*, 32 (4): 34–62. doi:10.1007/BF03374271.

Lister, M and L. Wells (2001), 'Seeing Beyond Belief: Cultural Studies as an Approach to Analysing the Visual', in T. van Leeuwen and C. Jewitt (eds), *Handbook of Visual Analysis*, 61–92, London: SAGE Publications.

Mah, A. (2009), 'Devastation but also Home: Place Attachment in Areas of Industrial Decline', *Home Cultures*, 6 (3): 287–310. doi:10.2752/174063109X12462745321462.

Martínez, F. and P. Laviolette (2016), 'Trespass into the Liminal: Urban Exploration in Estonia', *Anthropological Journal of European Cultures*, 25 (2): 1–24. doi: 10.3167/ajec.2016.250201.

Marx, K. and F. Engels (1975), 'The German Ideology', in R. C. Tucker (ed.), *The Marx-Engels Reader*, 146–202, New York: Norton.

Matila, T. (2021), 'White death: Finnish World War Two narrative and alternative heritage work in social media', *International Journal of Heritage Studies: IJHS*, 27 (6): 617–34. doi:10.1080/13527258.2020.1843520.

Matila, T. (2022), *Seeing the War through a Finnish Lens: Representation and Affect in the World War II Photographic Heritage*, Oulu: Oulun yliopisto.

Matila, T., Äikäs, T., Herva, V.-P., Seitsonen, O., and Koskinen-Koivisto, E. *manuscript*, 'Smokestack Nostalgia or Hypermodern Spirituality? Urban Exploration in Abandoned Factories'.

Melin, H. (2020), 'Finland Is Still a Class Society', *Alusta, Tampereen yliopiston yhteiskuntatieteiden tiedekunnan verkkojulkaisu*, translated by V.-P. Ketola, available at https://www.tuni.fi/alustalehti/2020/02/21/finland-is-still-a-class-society/.

Meyer, J. F. (2001), *Myths in Stone: Religious Dimensions of Washington, D.C.*, Berkeley: University of California Press.

Monick, E. (1987), *Phallos: Sacred Image of the Masculine*, Toronto: Inner City Books.

Murtorinne, E. (1995), *Suomen kirkon historia: 4, Sortovuosista nykypäiviin: 1900–1990*, Porvoo – Helsinki – Juva: WSOY.

Näre, S. (2004), *Kokonainen nainen*, Helsinki: Kirjapaja.

Numminen, P. (2020), 'Siivooja Alma löysi 5 euron setelin ja palautti sen italialaiselle omistajalleen – "Suomi lopeta! Tämä on liian paljon jopa sinulta"', *Iltalehti*, 26 January 2020.

Nurmi, T. (2022), 'Näsinneula on menettämässä Suomen korkeimman rakennuksen paikan, pitäisikö mastoa pidentää 18 metrillä? – "Loistava ehdotus"'. *Tamperelainen*, 8 October 2022.

Orange, H. (ed.) (2016), *Reanimating Industrial Spaces: Conducting Memory Work in Post-industrial Societies*, London: Routledge, Taylor & Francis Group.

Parente-Čapková, V., Grönstrand, H., Hapuli, R. and Launis, K. (eds) (2015), *Nainen kulttuurissa, kulttuuri naisessa*, Turku: Turun yliopisto.

Peckham, R. (2009), 'The city of knowledge: rethinking the history of science and urban planning', *Planning Perspectives*, 24 (4): 521–34.

Piff, P. K., P. Dietze, M. Feinberg, D. M. Stancato and D. Keltner (2015), 'Awe, the Small Self, and Prosocial Behavior', *Journal of Personality and Social Psychology*, 108 (6): 883–99. doi:10.1037/pspi0000018.

Preston, J. L. and F. Shin (2017), 'Spiritual experiences evoke awe through the small self in both religious and non-religious individuals', *Journal of Experimental Social Psychology*, 70: 212–21. doi:10.1016/j.jesp.2016.11.006.

Pullinen, J. (2023), 'Sanna Marin kohtasi aitoa naisvihaa, mutta moni vierasti ja ihaili hänessä jotain ihan muuta', *Helsingin Sanomat*, 16 April 2023. Available at https://www.hs.fi/mielipide/art-2000009515094.html.

Rautio, J. and Kohvakka-Viinanen, A. (2009), *Tampere*, Ranua: Mäntykustannus.

Reading, A. (2015), 'Making Feminist Heritage Work: Gender and Heritage', in E. Waterton and S. Watson (eds), *The Palgrave Handbook of Contemporary Heritage Research*, 397–410, Basingstoke: Palgrave Macmillan.

Richter, R. (2017), 'Industrial Heritage in Urban Imaginaries and City Images', *The Public Historian*, 39 (4): 65–84.

Rimpiläinen, A. (2019), 'Nuottasaaren piipun purkamisesta käytiin laajat neuvottelut – rakennuksia ei ole lailla eikä kaavalla suojeltu', *Kaleva*, 5 March 2019. Available at https://www.kaleva.fi/nuottasaaren-piipun-purkamisesta-kaytiin-laajat-ne/1729870.

Sauer, B. (2014), '"Speedy Cars, Perky Women, Champagne and Striptease Bars": Neoliberal Masculinity in Crisis?', in Kathleen Starck (ed.), *A Man's World? Political*

Masculinities in Literature and Culture, 81–96, Cambridge: Cambridge Scholars Publishing.

Silberman, N. (2007), 'Chasing the Unicorn? The quest for "essence" in Digital Heritage', in Y. E. Kalay, T. Kvan and J. Afflek (eds), *New Heritage: New Media and Cultural Heritage,* 81–91, London: Routledge, Taylor & Francis Group.

Siltala, J. (1994), *Miehen kunnia: Modernin miehen taistelu häpeää vastaan,* Helsinki: Otava.

Suova, M. (1942), *Das ist Suomi: Finnland in Bild und Wort = Suomi kuvina*, Porvoo: WSOY.

Taksa, L. (2020), '"Hidden in Plain Sight": Uncovering the Gendered Heritage of an Industrial Landscape', in S. De Nardi, H. Orange, S. C. High and E. Koskinen-Koivisto (eds), *The Routledge Handbook of Memory and Place*, 204–13, Abington: Routledge, Taylor & Francis Group.

Talvi, J., M. Salmelainen, U. Wager, P. Haltia and E. Mainiemi (1972), *Tunne Suomi = Lär känna Finland = Learn to know Finland = Kenne Finland: 800 värikuvaa = färgbilder = colour pictures = Farbfotos*, Helsinki: Kuultokuva.

Taylor, S. (2010), *Narratives of Identity and Place*, London and New York: Routledge.

Wajcman, J. (1991), *Feminism Confronts Technology*, North Sydney: Allen & Unwin.

Wurst, L. and S. A. Mrozowski (2016), 'Capitalism in Motion', *Historical Archaeology,* 50 (3): 81–99. doi:10.1007/BF03377335.

5

Co-Constructing Chernobyl: Investigating Visitors' Heritage Meaning-Making through Performativity in the Chernobyl Exclusion Zone

Veera Ojala

Introduction

The nuclear power plant accident in Chernobyl on 26 April 1986, was one of the largest catastrophes in recorded history, and the worst-ever disaster in the history of nuclear reactors (Medvedev, 1990; Chernousenko, 1991; Higginbotham, 2019). Thirty-one people died as an immediate result of the explosion, and the contamination forced the evacuation and resettlement of 350,000 people. More than 600,000 so-called liquidators were involved in securing and cleaning operations in the area. Current estimates place the mortality toll somewhere between 4,000 deaths estimated by the United Nations and 90,000 suggested by Greenpeace. The full consequences of the disaster, including the long-term social consequences, are still being debated and studied. Many controversies, the accident's unprecedented nature and the unknown nature of its consequences have all contributed to its symbolic power (Dobraszczyk, 2010).

In 2011, the Chernobyl Exclusion Zone (CEZ) was officially opened for tourists, and before Russia's attack on Ukraine the number of visitors grew steadily. The zone has gradually been marketed, and the potential of the contaminated area was expected to be utilized to the fullest in tourism as the tour providers and the administrative personnel were launching the process of seeking UNESCO heritage status for the area in 2020. Had this goal been achieved, the zone would have joined the list of other culturally significant and infamous milestones regarding the history of nuclear power, alongside the Hiroshima Peace Monument, the Castle Bravo launching platform at Bikini

Atoll and the Erzgebirge uranium mine region on both sides of the German–Czech border. This emerging heritage branch, also designated as nuclear cultural heritage (Rindzevičiūtė, 2021), investigates the meanings of nuclear industry for societies and communities, and pursues to elaborate its sociocultural impacts. The global Covid-19 pandemic and the Russian attack on Ukraine interrupted the evolving processes of heritagization, as those were taking place before 2020, and to this date the future of the exclusion zone remains open.

Because of the disaster's long-term and far-reaching social, cultural and economic effects, it has not remained unnoticed in the social sciences or humanities. Along with detailed accounts of the causes and results of the explosion at the power plant (Medvedev, 1990; Chernousenko, 1991; Plokhy, 2018; Higginbotham, 2019), a branch of studies has investigated the cultural significance of the zone, mainly under the concept of 'dark tourism'. The major criticism of the term 'dark tourism' notes that the concept simplifies the experiences of the visitors, and recently emerging scholarship has studied the CEZ under the umbrella of heritage studies through various approaches (Hryhorczuk, 2019; Banaszkiewicz, 2022; 2023; Ojala, 2022; Duda, 2023). These studies valorize, for example, the processes of heritagization and the role of tour guides as mediators of experience and communicators of the value of heritage resources (Hannam and Yankovska, 2013; Rush-Cooper, 2020; Banaszkiewicz, 2023).

The studies that scrutinize tourists' activities in the zone have been marginal; until recently, when visitors' embodied responses, interpretation processes and engagements in the CEZ have acquired more academic attention (Duda, 2023; Farkić, 2020; Farkić and Kennell, 2021; Hryhorczuk, 2019; Ojala, 2022). This current study seeks to advance knowledge of different types of visitors in the zone by emphasizing their spatial practices while visiting the site and the ways these performative practices mutually co-produce and order the tourismscape (Van der Duim, 2007) of Chernobyl. Further contributing to the emerging field of heritage studies through the lens of visitors' spatial strategies, the three identified performative patterns in this study will be attached to the framework of ambivalent heritage, and visitors' spatial strategies will be analysed as practices that actively contribute to the recomposition of the zone's memory space.

Therefore, instead of solely emphasizing the difficult and negative aspects of heritage, this study brings forth varying ways to engage with heritage resources and the processual character of heritage as entangled and intertwined through various stakeholders (Storm, 2014). Elaborating on the spatial strategies of different visitors to the zone reveals conflicting attitudes on how to use and preserve the site, reminding us that heritage-making is an ongoing process, constructed and challenged by visitors' embodied on-site performances. This study builds further on the studies that have concentrated on the various relationships with heritage in the CEZ (Ojala, 2022; Banaszkiewicz, 2023) and how the spatial usages of space interact with divergent memory representations in a developing heritage site. By absorbing a bottom-up approach supported by ethnographic fieldwork and semi-structured interviews with various visitors to the zone, this study provides a more nuanced picture of the activities of different visitors and how they contribute to the mediation and construction of tourismscape through visitors' performative practice regimes and identity hierarchies.

Theoretical framework

This study seeks to advance the studies about tourism in the CEZ from a heritage perspective under the concept of ambivalent heritage (Chadha, 2006; Storm, 2014; Donaldson, 2018), particularly highlighting two important aspects of the phenomenon: the international character of tourism in the CEZ and the various spatial strategies of the visitors to the zone. The zone, without question, is a signifier of a human and ecological disaster, and for countless individuals the aftermath of the disaster is still an ongoing trauma brought about by evacuation and health impairments (Yaroshinskaya, 2011). Visitors to the zone are aware of the heavy legacy of the nuclear accident, but among the visitors, various motivations can be identified for visiting the site, where commemoration can be one factor. As Banaszkiewicz (2023: 94) also acknowledges: 'the heritage of the CEZ turns out to be too heterogeneous to classify it as a solely dark site, and the experience of exploring the zone escapes the simple description of fascination with death.' In this article, I will examine various visitor categories, their diverse sensory experiences, their spatial practices, and how these turn into various and ambivalent heritage interpretations. In the context of gradual marketization, various heritage discourses occur simultaneously, along with the authorized discourse of the heritage and a range of popular discourses and practices (Smith, 2006), in addition to the visitors' identity hierarchies.

The ruins of Pripyat provide a scenic counter-reality that is encountered through various sensory stimuli. The visitor may testify to the victory of nature over man-made ideologies and infrastructures, and reflect on the transient nature of human lives (Farkić and Kennell, 2021). The exclusion zone escapes straightforward heritage interpretation as its imaginaries and representations are many and varied. The CEZ and its various divergent remains, the ruined city of Pripyat, abandoned villages and the power plant, belong also to industrial heritage (Storm, 2014; Banaszkiewicz, 2023) which leaves a legacy that signifies sorrow, the abuse of power and environmental pollution but also conveys stories on industrial communities and the achievements of technology and mankind. Storm (2014) characterizes these ambiguous and complex pasts that carry the meanings of fear, survival, resilience and courage as landscape scars, with divergent intertwined experiences combined with pain and brighter attributes. The city of Pripyat, the Soviet utopia Atomgrad and its contemporary ruins are primary examples of these types of ambivalent heritage resources, which are interpreted through individual meaning-making processes and spatial strategies, and evoke strong sensory experiences, such as nostalgia, sublime, pain and discomfort, in addition to the joy and mystery of the atmospheres of abandoned ruins.

Even though the interpretation of the zone's past can be situated within the framework of difficult and dissonant heritage (Tunbridge and Ashworth, 1996), the way various tourists create relationships with the heritage resources requires further elaboration. In addition to the essential ambiguity of the zone's toxic heritage, the approaches to the materiality of the zone are framed by divergent styles in a dynamically evolving tourism context. The concept of ambivalent heritage disrupts the dichotomic perceptions of heritage as either-or, rather emphasizing the process, the ambiguous character of the materiality and its emotional and personal content and interpretation (Chadha, 2006: 358). Rejecting the clear-cut understanding of temporality and change, and searching for a holistic perception of the processes of interpretation and construction of the experience through various stakeholders, the ambivalent heritage of the zone will be analysed by acknowledging 'its past significances, which are entangled with present standpoints' (Storm, 2014: 3). With this chapter, I exemplify how individual, personal engagements with the material remnants of the past in the zone, in combination with the visitors' agency and spatial strategies, influence the making of the place – the zone – in profound ways.

As a study that would close-read visitors' spatial strategies and how those contribute to the overall dynamics of the tourismscape is still missing, this present study aims to fill the research gap by utilizing a performative approach that assumes the presence and activity of tourists as active agents through their spatial strategies (Smith, 2006; Molnár et al., 2023). As Banaszkiewicz (2023: 94) alleviates, 'a key observation made by dark tourism researchers in the context of guides' narratives is the perception of the variety of tourist experiences in places of dissonant heritage, which is related to the complexity of tourist motivations and needs'. With the performative approach (Smith, 2006), this study sheds light on the visitors' spatial strategies and the ways those are situated in the wider transnational tourismscape, understanding the role of the visitors in a tourism setting as producing both conformity and friction (Pfoser and Keightley, 2021). The visitors negotiate their understanding and experiences of heritage in the context of gradual marketization, within the interplay of official and unofficial spatial practices and other visitor subjectivities. Like processes of place-making, understandings of ambivalent heritage are constantly being remade by diverse new experiences and emotions. Thus, heritage places are not only representative of past experiences, but also influence current experiences and perceptions of the world (Smith, 2006: 77). Meanings of heritage emerge through contemporary interactions within physical heritage sites.

The zone was, in the year 2019 when the material for this study was collected, a dynamically evolving tourism site where the process of gradual marketization was realized in the number of visitors, diversity of different tours and improvement in the visitors' infrastructure (Banaszkiewicz et al., 2017: Ojala, 2022). The year 2019 was also a record year, with 124,000 visitors. Therefore, the chosen time also enables the study of the influence of the gradual standardization and marketization of the heritage on the visitors' spatial practices. Following the understanding of the ambivalent heritage and the perception of the exclusion zone as a scar (Storm, 2014) constructed through varying voices, this study approaches the zone as a transnational tourismscape, first by identifying divergent visitor categories and then elaborating on the internal dynamics of these categories.

In the context of tourism in Chernobyl, the strategies visitors use to resist the official discourses (Gutman and Wüstenberg, 2022; Molnár et al., 2023) of the tour providers and the processes of marketization are an under-investigated approach. Nevertheless, 'heritage is defined and redefined by the present-day articulation and negotiation of values connected to stories, places, actions and events' (Storm, 2014: 5). Heritage tourism in this context is an umbrella term for a diverse set of mnemonic activities that can be seen as spatial strategies mobilized by visitors in their desire to seek reconciliation independent of authorities and official narratives (Molnár et al., 2023: 383). Turning the lens towards the corporeal proximity to the materiality of the heritage site enables us to understand places as practiced, produced and performed (Bærenholdt et al., 2004) and, importantly, challenged. Taking the processes of conformity and friction into consideration, in addition to the diversity of the visitors' profiles in the zone, this study provides fresh insights into the engagements with the ambivalent heritage from the bottom-up visitors' perspective and how the meaning of the heritage is negotiated.

The tourism industry in the CEZ has developed in phases. Since 2002, it has been possible, in practice, to visit the zone, and the zone was officially opened for tourists in 2011. Unofficial tourism has also developed alongside it (Banaszkiewicz, 2022; Ojala, 2022). In the context of the CEZ, unofficial tourism has taken place since the mid-nineties, and in 2017–19, before the Covid-19 pandemic and subsequent Russian attack on Ukraine, the unofficial tours also became commodified, and various representations such as visual material on social media, YouTube videos and books started to circulate, signifying the phenomenon (Stepanets, 2017; Richter, 2020; Kamysh, 2023). This study contextualized these different tour possibilities, which work on both sides of tourism – official and unofficial – as spatial strategies and performances that co-create order on the tourismscape (Van der Duim, 2007) in a shared social and cultural sphere, and these categories must be understood through their interconnections. As Banaszkiewicz (2022: 87) notes: 'It is beneficial to understand legal and illegal tourism as a spectrum of behaviours that inspire each other.' This study draws attention to these practices of commemoration and further expands the horizon of analysis, starting from the conclusion that spaces of ambivalent heritage are continuously renegotiated and reconstructed.

Methods and data

This study was conducted with a mixed-methods approach, combining an ethnographic case study approach with qualitative interviews, particularly concentrating on the demand side of tourism (Cakar and Aykol, 2021). The examination of the context started as fieldwork, which took place in Ukraine over two months in September and October 2019 (Ojala, 2020). I also visited the CEZ with two different types of tours, and this engagement with the tours and the overall fieldwork in Ukraine provided a useful overview of the types of tours currently available in the area. The tours offered a different level of structured information along with highly distinct modes of engagement with other tourists and the landscapes of the exclusion zone. I paid attention to the construction of the tour, the reactions of other visitors, and the type of spatial strategies they utilized in the area through their movements and levels of engagement. To meet the diverse set of visitors and their needs, in the year 2019, there were a plethora of different tours available, ranging from a one-day tour to multiple-day private tours that can be fully customized to a client's needs. For visitors who are after an intense experience, it is possible to book online the so-called unofficial or 'Stalker tour' into the exclusion zone. The latter presents itself as an alternative to the standardized 'mass-market tours' (UrbexTour, 2022). My engagement with the tours and the overall fieldwork in Ukraine provided context knowledge that was further used when continuing the interviews with the actual interest group of this research, the visitors to the exclusion zone.

The context research was further supplemented with semi-structured interviews with the interest group of this study: the individuals who visited the CEZ in 2019. As the topics of this study comprise individual meaning-making processes, the most suitable way to approach such an interpretative sphere is to attend an elaborative reciprocal discussion with the participants of the research. The interviewees were recruited from Facebook platforms in September 2019 and January 2020. The establishment of these online communities is based on the idea of sharing various heritage experiences after visiting the CEZ. In these groups, the members usually publish their photographs and share their impressions from their visits. The group members were approached by sending private messages where I introduced myself and the purpose of the study.

In most cases, the research participants had visited the exclusion zone a while before, ranging from a few months to a few weeks, thus enabling them to have some time to reflect upon their experiences. This facilitated the study and allowed it to maintain its international focus. The social media platforms enabled communication with participants from seventeen different countries of origin. Their ages vary between twenty-one and fifty-nine, and they come from different occupational classes in society (Table 5.1). Although I sent interview requests

Table 5.1 The profile of qualitative interviews, n=40

Subject	Sex	Age	Origin	Type of tour*	Year of visit**	Number of visits	Occupation
Respondent 1	M	46	Sweden	Illegal	2017	4	B
Respondent 2	M	22	United Kingdom	Illegal	2019	1	CE
Respondent 3	M	46	Denmark	Private	2019	1	NS
Respondent 4	F	33	United Kingdom	Group	2016	1	P
Respondent 5	M	25	Belgium	Group	2019	1	B
Respondent 6	F	31	Ukraine	Illegal	2016	5	P
Respondent 7	M	52	United Kingdom	Private	2018	2	P
Respondent 8	M	33	United Kingdom	Illegal	2019	1	P

Continued

Table 5.1 *(Continued)*

Subject	Sex	Age	Origin	Type of tour*	Year of visit**	Number of visits	Occupation
Respondent 9	M	31	Italy	Group	2019	1	P
Respondent 10	M	53	United Kingdom	Private	2016	3	CE
Respondent 11	M	29	Belgium	Illegal	2019	1	CE
Respondent 12	M	34	Belgium	Private	2019	1	P
Respondent 13	M	28	Germany	Private	2019	1	P
Respondent 14	M	34	United Kingdom	Private	2019	1	CE
Respondent 15	F	43	United Kingdom	Group	2019	1	CE
Respondent 16	M	44	Austria	Private	2017	2	P
Respondent 17	M	34	France	Illegal	2019	2	CE
Respondent 18	F	38	Poland	Private	2018	3	P
Respondent 19	M	45	France	Private	2014	7	P
Respondent 20	M	42	Australia	Group	2016	1	P
Respondent 21	M	44	Belgium	Private	2017	2	CE
Respondent 22	M	49	United Kingdom	Group	2016	2	CE
Respondent 23	M	59	United States	Private	2019	1	NS
Respondent 24	M	25	Norway	Private	2019	1	CE
Respondent 25	M	44	Germany	Group	2019	1	NS
Respondent 26	M	38	Germany	Private	2012	10	CE
Respondent 27	M	42	Belgium	Private	2019	1	CE
Respondent 28	M	32	Germany	Group	2019	1	P
Respondent 29	M	52	Australia	Private	2019	1	CE
Respondent 30	M	26	United States	Private	2019	1	S
Respondent 31	M	29	United Kingdom	Private	2013	4	CE
Respondent 32	M	37	Germany	Private	2019	1	NS
Respondent 33	M	43	Portugal	Illegal	2019	1	NS
Respondent 34	F	38	Finland	Group	2019	1	P
Respondent 35	M	21	Sweden	Group	2019	1	S
Respondent 36	M	44	United Kingdom	Private	2019	1	CE
Respondent 37	F	30	Netherlands	Private	2019	1	P
Respondent 38	M	32	Finland	Group	2019	1	CE
Respondent 39	M	38	Czech	Group	2019	1	CE
Respondent 40	M	40	Poland	Private	2012	2	P

* If a respondent had visited the exclusion zone with several types of tours, the category equals the one that the respondent preferred.
** If a respondent had visited the exclusion zone more than once, the category equals the year of the first visit.
M= Male, F= Female.
B= Businessperson, CE= Company Employee, NS= National Service, P= Professional, S= Student.
Source: Author's data, Ojala, 2020.

equally to both genders, they were more commonly accepted by men, thus men are more represented. Thirty-eight out of forty were conducted with the assistance of Skype. Two other respondent interviews were conducted in September and October 2019 in Kyiv, Ukraine.

A total of forty (n=40) semi-structured, in-depth interviews ranging in length from 30 to 90 minutes were conducted in English. Interviews focused on the particular tour choices, visitors' experiential benefits, interactions during the visit and practices of exploration. The transnational perspective of this study provided rich research data. More importantly, it also allowed the emerging characteristics of different visitors' categories – group, private and unofficial – and their group-specific relationship to the material resources in CEZ to be visible. These observations were further processed into themes corresponding to different spatial strategies when conducting the data analysis.

All interviews were audio-recorded with the participant's permission and transcribed verbatim in English. I organized and coded the data using Atlas.ti software and further analysed it with narrative analysis (Table 5.2). Narrative analysis was chosen because it is appropriate for exploring intrapersonal and interpersonal participant experiences and actions to understand a specific phenomenon through a story (Saldaña, 2016). One of the main assumptions of narrative analysis is that people's stories are important lenses through which it is possible to explore social reality (Mura and Sharif, 2017). As the aim of this study was to acquire a holistic understanding of a given social and cultural sphere – the tourismscape of the CEZ at a chosen point in time in 2019 – narrative analysis was deemed an appropriate tool for the analysis as it targets scrutinizing 'the social phenomena that activate, produce, organize and transmit stories' (Mura and Sharif, 2016: 195). In other words, by analysing the way individuals construct and represent events in their lives, we can take a picture of social phenomena at the macro level, as narratives are 'situated' in a particular sociocultural context.

Table 5.2 The results of the thematic narrative coding

Type of tour	Group tour	Private tour	Unofficial tour
Performative disposition	Spectator	Explorer	Adventurer
Spatial strategy	Conformity	Enactment	Contestation
Risk perception	High	Moderate	None

Source: Author's data, Ojala, 2020.

In narrative analysis, the focus is on 'the stories that people employ to account for events' (Bryman, 2016: 590). The narrative analysis was used as a means of unpicking the details within a given account of a participant (Grbich, 2013: 216), and the deeper meaning of the details in the used language and expressions were further processed into themes and sub-themes. First, I identified the tour categories' particular performative dispositions: spectator, explorer and adventurer, as it was decoded from the transcripts, and it corresponds with the sought level of engagement with the materiality of the zone. Second, I linked each disposition to distinct spatial strategies: conformity, enactment and contestation that belonged to the particular activities of the tour category and corresponded to the visitors' mutual dynamics that comprised the tourismscape in 2019. The perception of the risks of radiation was also acknowledged, as it adds a further layer to the usage of space and the interpretation of the ambivalent heritage resources in the zone. Next, the results part represents these spatial strategies and approaches to the heritage resources in the zone and their overall dynamics in the tourismscape.

Performing the zone as a spectator

This study identified three performative patterns in the CEZ (Figure 5.1). These performative traces were a combination of visitor-specific performative dispositions, which comprised category-specific spatial strategies in combination with the risk perception regarding the radiation in the zone. The first performative

Figure 5.1 Pripyat town sign, completed 1970 (Photo: Veera Ojala, 2019).

pattern was produced by the visitors from the group tour category because of their specific ways of engaging with the materiality of the zone.

When visiting the CEZ with a group tour, visitors follow the route itinerary that consists of the main sights in the exclusion zone, such as Duga radar, Chernobyl town, the nuclear power plant and the abandoned city Pripyat. The group size usually varies between 10 and 25 people, and visitors are accompanied by the tour guide the whole time. Tour guides structure the visit by providing information on each location visited, not only regarding the history of the site but informing about the radiation levels, which are demonstrated in some places with the assistance of a dosimeter. Starting in the early morning in Kyiv, the tour buses arrive at the exclusion zone a couple of hours later, and after passing the main checkpoint Dytyatky, where the verification of the identity of the visitors takes place and they receive a personal dosimeter, the buses scroll into the exclusion zone. One respondent (15) described her decision to visit the exclusion zone:

> That given its change nature will find the way and I wanted to go and visit the abandoned town. And see the fact that nature has started to recuperate things and living really, it is the sort of going around a ghost town, that's the main reason.

The spontaneous decision to visit the exclusion zone and the activity of 'going around the ghost town' describe incisively the group tour visitors' specific style of travel to the exclusion zone and the relationship with the landscapes they sought to experience. Approaching the CEZ within a performative disposition of *spectator*, the respondents did not seek to create a specifically pronounced engagement with the material resources in the zone. They simply wished to see and visit the place. Spatial practices of the spectators were mostly concentrated on the exterior surfaces in the exclusion zone. They recognized nature and its spatial outcomes in the area. Equally, they observed the built environment, its decay and the disappearance of its silhouettes under the vegetation, which was a powerful sensory encounter.

> The abandoned villages [. . .] that's really frightening to see how people used to live over there, how they used to have a life, with all kind of buildings and all kind of facilities and see how it is all now abandoned and destroyed. That's really frightening, that was most impressive of all of it.
>
> Respondent 5

The witnessing of exterior surfaces did not only have an impact on the respondents because of the loss that they were reminded of, but also because the landscapes signified toxic otherness:

> I remember one point when the ground started to smell like a really nice scent of fresh ground, and I thought that I cannot breathe because I don't know if I will be breathing some particles. But I think you manage in that world if you just stick with the rules.
>
> Respondent 34

What also informed the spatial strategies of the spectators was their awareness of the radiation risks. A common denominator in their spatial strategies was that a sufficient level of engagement with the material spaces in the zone could be characterized as a stance of distant contemplation. Spectators' estimation of the risks of being exposed to radiation or to the dangers of decaying infrastructure informed their spatial movements, which in turn influenced their particular performance in the zone.

Therefore, they sought to conform to the safety rules, pre-determined routes and landscapes, not to outweigh them by entering them. These spatial strategies, therefore, focused on outside explorations and observations of the environment on the marked routes. In consequence, their practices of co-composing the tourismscape were formed by moving through its exterior surfaces and being impressed by what they saw. The decaying silhouettes of the city and villages were 'impressive' to look at, but they contained dangers and risks, and looking at them from the position of a distant contemplation was a sufficient level of engagement for these respondents. Through the spatial strategies, focused on the exterior surfaces and engagement of distant contemplation, the spectators can be perceived as a performance trace in the tourismscape that will be named *conformity.*

Performing the zone as an explorer

The maximum number of individuals who participate in the private tour is smaller than in the group tours, and the visitors have the possibility to choose which locations they wish to see and have an impact on the time spent in each location. It allows for less restricted possibilities for spatial exploration of landscapes, buildings and artefacts that the ruins of the zone encase. Be they first-timers or multiple-timers, respondents were keen on photographs, influenced by urban exploration, the history of the site or otherwise curious to experience the atmospheric sides of the exclusion zone (Figure 5.2).

The second visitor group that emerged from this study and contributed to the multivocality of the ambivalent heritage site (Frihammar and Silverman, 2018),

Figure 5.2 Azure swimming pool in Pripyat (Photo: Veera Ojala, 2019).

were respondents who visited the exclusion zone with a private tour. Respondents chose a private tour because they wished for a deeper and more personal engagement with the spaces of the exclusion zone. The private tour also appealed more to them because of its customized and personalized character. A common denominator in the experience of visitors from the private tour category was that they were less satisfied with standardized, regulated tourist activities in the area. A respondent (29) reasoned his decision to take a private tour with the following argument:

> Tourists and buses were not what we were looking for, we wanted to explore, where we wanted to, as long as we wanted to, we did not want to be among other tourists, we wanted to explore buildings and artefacts completely on our own.
>
> Respondent 29

Personal engagement, the possibility to actively shape one's sensory experiences, mastering one's time plan and the freedom for spatial exploration, were those determining experiential aspects that respondents sought from the private tour. The performative disposition to the zone describes the spatial strategies of the private tour category, referred to as the *explorers*.

Also visiting the same locations as the group tour, the private tour enables the possibility of visiting more unknown locations in Pripyat and the wider exclusion zone. In addition to the possibility to visit the 'sites that were not touristic', a pronounced engagement in the zone among the explorers were the artefacts and the interiors of the decaying buildings. The explorers' choice of the tour was characterized by the desire to shape autonomous sensory experiences in the zone. The experiences from the inside of the ruins were a combination of sensory reactions because of various stimuli, the absence of life, the lack of sound and the amazement for the objects that could still be found inside, which easily create a sense of overwhelming because of the excess of meaning generated by the recognition of familiarity in unfamiliarity. Because of the ambiguous and embodied sensory encounters, the inside exploration of the houses is a prime example of the type of ambivalent personal engagement with the heritage resources through which the respondents sought to enact their experiences. The interviewees sought to create an intimate relationship with the ruins and decaying structures of the exclusion zone.

> The place is abandoned for the most part, 50,000 people, so being there with a big group you miss the soul of this place. You miss being inside a building because the ten of you can't go inside the buildings and just sit there and listen to the wood creak from the window outside. You miss that if you are in a big, noisy, loud group.
>
> Respondent 14

The excerpt accurately describes the experience that explorers sought when visiting the exclusion zone. In addition, it exemplifies equally the distinct spatial strategies of explorers: they wished to engage with the zone through personal experiences that echoed the embodied, sensuous and felt sides of the zone. The ambiguity was acknowledged by respondents as a contributing factor to the establishment of a possibility for more intimate engagement with those material remains that now signify the disaster. The risks of radiation were also acknowledged but bargained for between the acquisition of 'real' and personal experience.

> And I went to the roof, and I rested there for 10–15 minutes, just walked over the roof which scared me a little bit because I did not have a dosimeter with me. I don't know if people go there but on the other hand, I had a really quiet and personal moment there.
>
> Respondent 13

The type of trace that explorers left on the tourismscape is called *enactment*. This enactment was because of the deeper personal engagement they sought; they wished to enact their possibilities for material engagement by choosing a private tour, which would essentially give them more freedom for exploration and unrestricted sensory engagement with the zone (Figure 5.3). As this analysis discovered, the forms of enactment were the construction of the zone through atmospheric, embodied experiences by pursuing deeper proximity with the

Figure 5.3 Palace of Culture Energetik in Pripyat (Photo: Veera Ojala, 2019).

materiality of the zone than just distant contemplation. The decision of the spatial strategies was taking place as a dialogue between the personal interpretation of the heritage resources, individual negotiation of the risks of radiation and personal freedom. Therefore, the descriptive term for the visitors' performance is enactment of the zone, as they sought tangible and intangible affordances to enact their engagements with the heritage resources.

Performing the zone as an adventurer

When one wishes to visit the exclusion zone with so-called 'unofficial tours' or 'Stalker tours', the visitors enter the exclusion zone with their guide by passing underneath the barbed wire and hiking through the exclusion zone towards its epicentre, Pripyat, and back. Because of the necessity to stay unnoticed, unofficial visitors move mostly at night. The length of the hike is around 60 kilometres, and the usual duration of the unofficial trip is four days. As the name of the tour already hints, the unofficial tour is not obligated to follow the official safety regulations of the zone. Instead of negotiating the spatial strategies with the existing safety policies, unofficial visitors must be aware of the security personnel of the zone and official tours.

Visitors who enter the exclusion zone with the unofficial tour are often influenced by urban exploration; thus, abandoned landscapes are a highly appreciated destination for individuals who enjoy the sensations, ambivalence and atmospheres of abandoned ruins. Because unofficial tourism functions outside the framework of official tourism, unofficial tours form discursive opponents for mainstream tours, as well as promising the possibility to be 'a Chernobyl explorer and discover new life experiences instead of the mass market of plain and boring tours' (UrbexTour, 2020).

Respondent (2) described his motivations for the unofficial trip and the rejection of the framework of official tourism in the following way:

> It is scripted, there is never any excitement (official tours), I want to go the extra mile to gain experience and do what I want to do rather than have a scheduled route and obviously just go with them. I want to explore what I want to in my own time and see inside the buildings, look at the history, and experience it all instead of spending a day there and going home at night.

Freedom for spatial exploration was a pronounced desire among the unofficial visitors. Another respondent (17) continued:

> During the unofficial tour you can do what you want, there are frequently policemen and soldiers who walk around the buildings you are visiting, so you have to be fast and act in silence but that's all.

Thus, physical proximity to the landscapes and a desire for a unique, unrestricted encounter with the zone were pronounced articulations of engagement among respondents from this visitor category. Because of the corporeal proximation with the materiality of the zone and visitors placing themselves outside the established safety regulations of the zone, the narrative analysis found a common attribute for the spatial strategies of the visitor category; the third visitors' performative disposition this study identified is further named *adventurers*.

The unofficial spatial strategies that adventurers were adopting comprised divergent types of experiential dynamics. A respondent (33) described the most memorable part of his journey in the following way:

> The landscape and the road to Pripyat because you see a lot of nature and animals. And one interesting thing is the lack of people and nature, it is like going to another planet. Because sometimes you walk a long time without seeing anyone and you do not see any traces of human presence. You do not see houses, you do not see roads, it is different from day-to-day life where you are always surrounded by people and cars and buildings. It is a different experience.

In addition to personal engagement with the area, the respondents sought the possibility of unrestricted spatial practices as they moved through the landscapes. The close physical proximity to the zone, special movements through the area at night and the necessity to stay undetected were the patterns of exploration that characterized respondents' experiences who were part of the adventurer category.

> You feel that you are alone in the world, in legal tours, you can be there for a few hours and you always have people around, but if you go across the zone and see the small abandoned villages and you go to the forest through the countryside, you see a lot more and the quietness; you have the feeling that you are the last people on Earth when you walk in there. It is like the apocalypse happened; that feeling is quite wonderful because it is not crowded with people.
>
> Respondent 11

In addition, their particular spatial practices at night and outside the official routes added a layer of additional excitement. The adventurers had intense emotional engagements with the material resources in the zone through their specific style of exploration. These excerpts equally describe alternative uses of the space and exemplify the practices that contested the tourismscape by consisting of innovative ways of exploring the zone outside official routes.

> My favourite part was the night we came to Pripyat: we came 25 kilometres hiking, and the last 10 kilometres were very difficult. We had a lot of stress because you have to take the official road to go to Pripyat. The problem is that the road is occupied by the cars of Chernobyl workers and policemen. So, when you see lights, you have to jump in the bushes – it was, like in the movie, a very stressful moment. When I finished the trip to Pripyat, I was on the roof of a building and saw the dark night and reactor 4 lying in the night. it was a beautiful moment with a bottle of beer.
>
> Respondent 17

As this excerpt illustrates, the adventures' spatial strategies formed opposite statements for official tourism, thus contesting the tourismscape. It was not construed as an area to be afraid of or to approach cautiously and observe from a distance. Rather, it was a place to create an intimate relationship, which gave a sense of mystery and adventure to the lives of the respondents. The disposition of adventurers, the corporeal proximity to the landscapes and the exploration of the tourismscape outside the official framework of tourism allowed peculiar visitors' spatial practices in the zone. Moreover, the adventures co-compose discursively the tourismscape by innovating new routes, addressing it with contradictory connotations and challenging the official tourism structures. Therefore, the type of performative order that the adventures leave on the zone's tourismscape through their spatial strategies and practices of exploration is going to be further named *contestation*.

Dynamics of the tourismscape

The three performative dispositions that this study identified were *spectator, explorer* and *adventurer.* These descriptive labels for each visitor category exemplify the depth of the interactive engagement with the materiality of the zone, and are in concordance with the three types of tours that visitors can attach themselves to when visiting the exclusion zone: group, private and unofficial tours. By following how meanings are created and distributed among people and things, and how these consist of interconnections, it is possible to identify the processes of ordering in heritage sites (Van der Duim, 2007). These performative patterns leave their traces on the tourismscape, through practices of *conformity, enactment* and *contestation.* Visitors engage with the heritage resources by following these consumer limitations and needs. They have specific spatial strategies through which they seek to either conform to or negotiate and exceed the existing framework of rules.

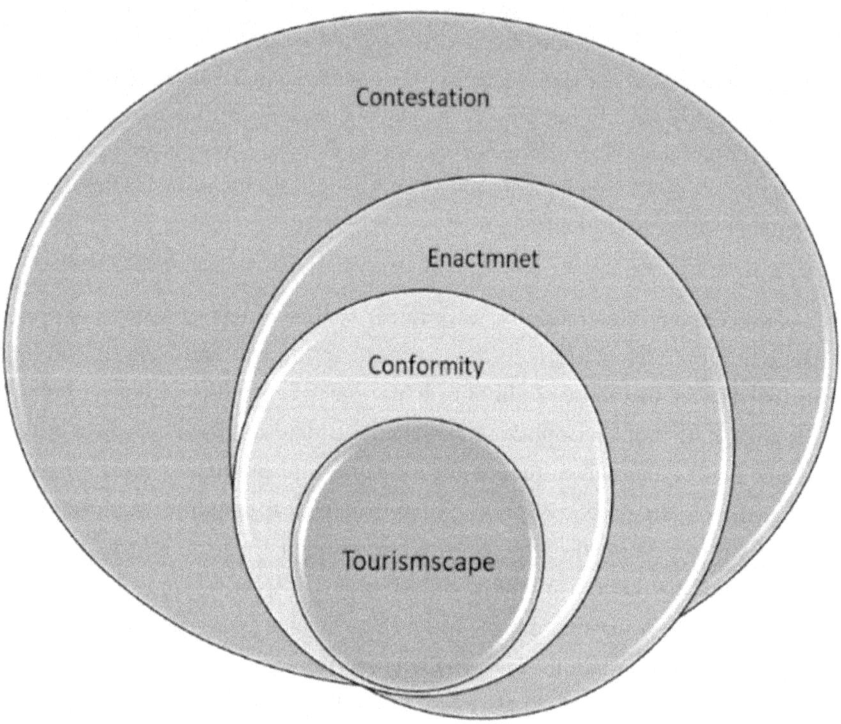

Figure 5.4 The performative ordering of the tourismscape in CEZ (Source: Ojala's data 2020).

Furthermore, although these differing visitor entities demonstrated differing degrees of engagement because of varying and ambivalent intensities in the material landscapes of the exclusion zone, there is a further tendency that these visitor entities can be interpreted as demonstrating.

> I was fearing that it had already gotten too big; my fear was that we would see too many people and that too many people would go there. The experience will not be so unique because, yeah, it is so overrun by people. But it did not happen; maybe it happened a little bit at the Ferris wheel.
>
> Respondent 13

These different tour approaches expressed satisfaction regarding the standardization of tourism. Visitors' performances exemplified the stakeholder's active agency and practices of producing a style of exploration that would be free from regulations (Edensor, 2001; Garret, 2015). Thus, there is a further dimension that these visitor entities stand for, as this excerpt reveals:

> It is not emotion or imagination (the official tours); the guide is just telling what this is and what that is. I want to dig into the story, the exact things that were left behind; that kind of thing is better explored by yourself in your own time to get a full picture of exactly what they went through.
>
> Respondent 2

What is informative in this excerpt is the tendency for tourists to always search for new places to go, places that would be 'untainted by the commercial emotional management of the tourism industry' (Brunsden and Goatcher, 2011: 131). The process of 'increasing the attraction factor' of attractions to appeal to new tourists hungry for stronger experiences (Banaszkiewicz, 2023: 92), can be verified with the typology of this study. Indeed, the desire for freedom for exploration and determination over one's sensory encounters can be seen as a crucial motivational factor for choosing a private tour or an unofficial tour.

Discussion

This study exemplified how tourists participate in the multivocality of the ambivalent heritage site through distinct performative patterns through which they consume particular aspects of the area. Tourism produces special heritage outcomes because tourists have an interest in special experiences, artefacts and narratives in the context of the heritage (Šešić and Mijatović, 2014). Particularly targeted to bring forth the spatial strategies of the three identified visitor categories and their divergent performative practices in the developing toxic and ambivalent heritage site, the study provided a comprehensive picture of the recomposition of the zone's tourismscape from a comparative perspective. As the varying choreographies and heritage practices that this study has elaborated exemplify, there is not only one way to engage with the heritage in Chernobyl. As Frihammar and Silverman (2018) characterize the sites of dissonance heritage as polysemous, which generate layers of conflicting emotion, memory and practice, the current study has indicated clear examples of how different tourist groups approach this ambivalent heritage, how they deal with this heritage, and furthermore, how the visitors' performative traces mutually reproduce and order the tourismscape.

As the narrative analysis found, the spatial performances of these different visitor categories reflected their different interests in the heritage. What these categories exemplify is the fact that the intensity of the sought-after experience varies between the visitor profiles. The typology of this study exemplifies a desire for determination over the sensory and embodied experiences of one's visitors' subjectivity. This again draws attention to the power and ambivalence of physical places in every tourist experience. The CEZ is a physical heritage site but also a cultural process that visitors use to express and construct a sense of identity in which the material site is invoked 'in its representational sense to give a physical reality to these expressions and experiences' (Smith, 2006: 75). The ambivalence in the heritage of Chernobyl depends on the reaction of people to that place and the intensity and nature of their experiences, through which they sought to construe their heritage's articulation.

This study was able to supplement previous research (Hannam and Yankovska, 2013; Rush-Cooper, 2020; Banaszkiewicz, 2023) conducted from the perspective of the tour guides by providing insight on performativity, embodiment and ambivalence of heritage that was grounded in the perspective of the respondents themselves. There is not a single type of tourist who enters the exclusion zone. As this study indicated, there are several patterns of performativity, and visitors engage and negotiate these patterns based on different tour choices. These visitor categories provide three different performative lenses into the exclusion zone and the interpretation of its socio-cultural meaning, which further supplements the discussion on the processes of ambivalent heritage.

The CEZ and the heritage it stands for do not create a single narrative authority. It is precisely because of this ambivalent heritage that varying interpretations and articulations exist in one place. These interpretations are the result of the varying levels of embodied engagements of *spectator*, *explorer* and *adventurer* with the materiality of the zone. Visitors construct the meaning of the heritage resources in the zone through these embodied interpretation processes, which are informed by the intensity of the experience and risk evaluation. The contribution of this study is that it could identify these different articulations, produced by visitors' spatial strategies, and exemplify the heritage-making patterns of the tourists by elaborating on their embodied sensory engagement with the heritage site they sought. The patterns could also be placed as the composition of the tourismscape, where the interconnectedness of different visitor profiles also becomes clear, and their motivations and activities reflect each other. By following how meanings and tasks are attributed to and distributed among people and things, it is possible to follow and elucidate the processes of ordering in heritage sites (Van der Duim, 2007).

References

Banaszkiewicz, M. (2022), 'The Transgressive Approach to the Dark Tourism Experience: Stalkerism in the Chernobyl Exclusion Zone', *Polish Sociological Review*, 219 (3).

Banaszkiewicz, M. (2023), *Tourism and Heritage in the Chornobyl Exclusion Zone*, London: Routledge, Taylor & Francis Group.

Banaszkiewicz, M. and A. Duda (2019), 'To be a S.T.A.L.K.E.R. On Architecture, Computer Games and Tourist Experience in the Chernobyl Exclusion Zone', in M. Gravari-Barbas, N. Graburn and J.-F. Staszak (eds), *Tourism Fictions, Simulacra and Virtualities*, 197–210, London: Taylor & Francis Group.

Banaszkiewicz, M., Z. Kruczek and A. Duda (2017), 'The Chernobyl Exclusion Zone as a Tourist Attraction: Reflections on the Turistification of the Zone', *Folia Turistica* (44): 145–69.

Bærenholdt, J. O. (2004), *Performing Tourist Places*, Aldershot: Ashgate.

Bryman, A. (2016), *Social Research Methods*, 5th edn, Italy: Oxford University Press.

Cakar, K. and S. Aykol (2021), 'Case Study as a Research Method in Hospitality and Tourism Research: A Systematic Literature Review (1974–2020)', *Cornell Hospitality Quarterly*, 62(1): 21–31. https://doi.org/10.1177/1938965520971281.

Chadha, A. (2006), 'Ambivalent Heritage: Between Affect and Ideology in a Colonial Cemetery', *Journal of Material Culture*, 11 (3): 339–63. https://doi.org/10.1177/1359183506068809.

Chernousenko, V. M. (1991), *Chernobyl: Insight from the Inside*, Berlin: Springer.

Dobraszczyk, P. (2010), 'Petrified Ruin: Chernobyl, Pripyat, and the Death of the City', *City* 14 (4): 370–89. https://doi.org/10.1080/13604813.2010.496190.

Donaldson, E. C. (2018), 'Place, Destabilized: Ambivalent Heritage, Community and Colonialism in the Marquesas Islands', *Oceania*, 88: 69–89. https://doi.org/10.1002/ocea.5182.

Duda, A. (2023), 'Tourist Photography and Heritage: The Case of the Abandoned City of Pripyat', *Journal of Heritage Tourism,* 18 (3): 384–400. doi: 10.1080/1743873X.2022.2161907.

Edensor, T. (2001), 'Performing Tourism, Staging Tourism: (Re)producing tourist space and practice', *Tourist Studies,* 1 (1): 59–81. doi: 10.1177/146879760100100104.

Farkić, J. (2020), 'Consuming Dystopic Places: What Answers are We Looking For?', *Tourism Management Perspectives*, 33: 1–8. doi: 10.1016/j.tmp.2019.100633.

Farkić, J. and J. Kennell (2021), 'Consuming Dark Sites via Street Art: Murals at Chernobyl', *Annals of Tourism Research*, 90: 1–13. doi: 10.1016/j.annals.2021.103256.

Frihammar, M. and H. Silverman (2018), *Heritage of Death: Landscapes of Emotion, Memory and Practice*, London: Routledge.

Garrett, B. (2015), 'Urban Exploration as Heritage Placemaking', in H. Orange (ed.) *Reanimating Industrial Spaces: Conducting Memory Work in Post-Industrial Societies,* 72–91, London and New York: Routledge.

Goatcher, J. and V. Brunsden (2011), 'Chernobyl and the Sublime Tourist', *Tourist Studies,* 11 (2): 115–37.

Grbich, C. (2013), *Qualitative Data Analysis: An Introduction*, 2nd edn, Croydon: Sage Publications.

Gutman, Y. and J. Wüstenberg (2022), 'Challenging the Meaning of the Past from Below: A Typology for Comparative Research on Memory Activists', *Memory Studies*, 15 (5): 1070–86. https://doi.org/10.1177/17506980211044696.

Higginbotham, A. (2019), *Midnight in Chernobyl: The Untold Story of the World's Greatest Nuclear Disaster*, New York: Simon & Schuster Paperbacks.

Hryhorczuk, N. (2019), 'Radioactive Heritage: The Universal Value of Chernobyl as a Dark Heritage Site', *Qualitative Inquiry,* 25 (9–10): 1047–55.

Kamysh M. (2023), *Stalking the Atomic City*, London: Pushkin Press.

Medvedev, Z. (1990), *The Legacy of Chernobyl*, New York: Norton.

Molnár, V., Koziura, K., and König-Paratore, F. (2023), 'Tours and Tourism', in: Y. Gutman and J. Wüstenberg (Eds.), *The Routledge Handbook of Memory Activism*, 381–4, Abingdon, Oxon: Routledge.

Mura, P. and S. P. Sharif (2017), 'Narrative analysis in tourism: a critical review', *Scandinavian Journal of Hospitality and Tourism*, 17 (2): 194–207. doi: 10.1080/15022250.2016.1227276.

Ojala, V. (2020), *Chernobyl Dreams: A Study on the Role of Narrative Imagination in Consumption Experience*, Arbeitspapiere 03/2020, Ost-Europa-Institut FU-Berlin. https://www.oei.fuberlin.de/soziologie/arbeitspapiere/index.htm

Ojala, V. (2022), '"Chernobyl Dreams: Investigating Visitors' Storytelling in the Chernobyl Exclusion Zone', *International Journal of Tourism Cities*, Vol. ahead-of-print No. ahead-of-print.

Pfoser, A. and E. Keightley (2021), 'Tourism and the Dynamics of Transnational Mnemonic Encounters', *Memory Studies*, 14 (2): 125–39. https://doi.org/10.1177/1750698019856059.

Plokhy, S. (2018), *Chernobyl: The History of a Nuclear Catastrophe*, New York: Basic Books.

Richter, D. (2020), *Chernobyl: A Stalkers' Guide*, FUEL Publishing.

Rindzevičiūtė, E. (2021), 'Nuclear Power as Cultural Heritage in Russia', *Slavic Review*, 80 (4): 839–62. doi:10.1017/slr.2022.1.

Rush-Cooper, N. (2020), 'Nuclear Landscape: Tourism, Embodiment and Exposure in the Chernobyl Zone', *Cultural Geographies*, 27 (2): 217–35. https://doi.org/10.1177/1474474019876616.

Saldaña, J. (2016), *The Coding Manual for Qualitative Researchers*, 3rd edn, Los Angeles, London, New Delhi, Singapore and Washington, DC: Sage.

Šešić, M. D. and L. R. Mijatović (2014), 'Balkan Dissonant Heritage Narratives (and Their Attractiveness) for Tourism', *American Journal of Tourism Management*, 3 (1B): 10–19.

Smith, L. (2006), *Uses of Heritage*, Abingdon: Routledge.

Stepanets, K. (2017), *Chernobyl Zone through the Eyes of a Stalker*, Kiev: SkyHorse.

Storm, A. (2014), *Post-Industrial Landscape Scars*, New York: Palgrave Macmillan.

Tunbridge, J. E. and G. J. Ashworth (1996), *Dissonant Heritage: The Management of the Past as a Resource in Conflict*, Chichester: J. Wiley.

UrbexTour (2022), 'Chernobyl Urbex Tour'. Available at www.urbextour.com/en/chernobyl-urbex-tour/ (accessed 14 May 2024).

Van der Duim, R. (2007), 'Tourismscapes an Actor-Network Perspective', *Annals of Tourism Research*, 34 (4): 961–76. https://doi.org/10.1016/j.annals.2007.05.008.

Yankovska, G. and K. Hannam (2013), 'Dark and Toxic Tourism in the Chernobyl Exclusion Zone', *Current Issues in Tourism*, 17 (10): 929–39.

Yaroshinskaya, A. A. (2011), *Chernobyl: Crime without Punishment, [25 years later]*, Brunswick: Transaction Publ.

6

Spiralling into a Labyrinth of Cultural Fantasies and Extractivism: Treasures, Extraordinary Undergrounds and the 'Temple of Lemminkäinen' (Sipoo, Finland)

Vesa-Pekka Herva, Oula Seitsonen, Tina Paphitis, Teresa Komu, Gabriel Moshenska and Risto Nurmi

Introduction

In late 2021, Finnish and international media published news about the search for a great treasure. As far away as New Zealand, a newspaper reported that 'A team of amateur treasure hunters in Finland claim to be on the brink of discovering the world's largest treasure trove, worth over $28.5 billion' (Bywater, 2021). The idea of this treasure, supposedly hidden in a vast cave under a peculiar rock-outcrop in Sipoo near Helsinki, was originally introduced in the 1980s, but it resurfaced in 2021 as an international team publicized their plans to continue the search for the treasure. Media reports of the search for 'the fabled treasure chamber that would make Tutankhamun's gold-laden tomb look like the last stall at a car boot sale' took a generally humorous tone, but there is an intriguing greater story – indeed something of a (fabricated) mythology, the 'Bock Family Saga' – behind the alleged treasure and the so-called 'Temple of Lemminkäinen' where it is supposedly hidden (Figure 6.1).

It may be unfair to describe the Bock Family Saga and the story of the hidden treasure as pseudohistory insofar as the story, or some of its elements, are not necessarily meant to be taken as literal descriptions of the past but might instead be intended as some kind of allegorical mytho-history. In any case, it is distinctively a creation of the Finnish mystic-eccentric Ior Bock, and has inspired the search for the treasure in the form of digging into and under the rock outcrop that is dubbed as the Temple of Lemminkäinen in the Bock Family Saga. Drawing

Figure 6.1 The entrance to the Temple of Lemminkäinen, with the members of the Temple Twelve team pumping the water out of the site in preparation for their excavations in 2022 (Photo: Oula Seitsonen, 2022).

from various mythologies, Bock's version of an alternative past is, in itself, a creative approach to heritage, whereas an industrial heritage comes to play more indirectly. Namely, long-standing cultural ideas about the weird and characteristically ambivalent subterranean realm with hidden treasures, as recounted by Bock, are also present in past and present extractive industries, although this is not as obvious as with the treasure in the Temple of Lemminkäinen (also cf. Herva et al., 2022). Thus, the latter enables putting on sharper relief some intangible heritage of human–underground relations with its various extraordinary dimensions that also mediate perceptions of industrial mining, albeit often implicitly and indirectly. And yet, identifying and understanding the significance of these ideas about the ambivalent and extraordinary underground realm is elementary to appreciating the cultural and emotional entanglements and cosmological implications of extractive industries, both past and present. This chapter takes something like a 'backdoor approach' and alternative take on that subject matter. We trace interlinks between various cultural ideas and practices, initially inspired by Ken Hiratsuka's piece of art at the site, which make us also encounter a host – indeed an entire labyrinth – of ambivalences and ambiguities between long-term (cultural) fantasies and imaginaries of the subterranean on the one hand and 'rational' ideas and affairs on the other.

Our labyrinth is made of 'alternative pasts', archaeology, treasures and extractive pursuits as forms of engaging with the underground. We seek to identify connections and a common ground between these different elements from the perspective of long and complex living legacies of the human perceptions of, and relationships with, the subterranean and how they intertwine dreams, fantasies, mythologies and histories. A basic idea that runs through this chapter is that for hundreds and thousands of years, the subterranean has been conceived of as a mysterious, alien, strange domain of the supernatural and of otherness, marvels, wonders and nightmares that are layered and entangled, as we and many others have detailed elsewhere (e.g. Komu, 2019; Herva et al., 2022). These ideas, concepts and perceptions, with roots deep in prehistory, are still present today, although their role and significance are often not recognized, and yet they can help to understand certain aspects of, and similarities between, different forms of engaging with the subterranean world, including (pseudo) archaeology and historical and contemporary extractive industries. Our mapping of these connections aims to show that apparent 'fringe' cases, such as the treasure hidden in the Temple of Lemminkäinen, actually echo broader cultural ideas and fantasies about the subterranean, and are relevant in various contexts (for broader discussions of pseudoarchaeology see Schadla-Hall, 2004; Derricourt, 2012; papers in Fagan, 2006; see Wilson, 2012 for an interesting comparison of subterranean pseudoarchaeology).

Myths, histories, imaginaries and materialities are intertwined in sometimes surprising and unexpected ways and in contexts – such as modern extractive industries – where one might not expect to encounter them. This is an important, albeit overlooked or unrecognized, aspect of industrial histories and industrial heritage, especially as regards understanding why extractive industries provoke such contradictory responses – that is, apocalyptic fear as much as dream-like high hopes of a better future (e.g. Komu, 2020). Extractive industries tend to be considered in very different terms from 'invented' legends and mythologies, such as the Temple of Lemminkäinen and its treasure, but they do have 'real' and 'fantastic' elements to them, at least partly because they are ultimately grounded in cultural perceptions of a mysterious subterranean with its ambiguous and ambivalent dimensions.

This is also why seemingly outlandish cases, such as the Temple of Lemminkäinen, can offer perspectives on how our lives are inextricably intertwined with fantasies, even in seemingly rational domains, such as technology and the economy. Digging deeper into ideas and narratives of alternative pasts can put the role and significance of cultural ideas and fantasies in other domains of contemporary culture and society in a sharper relief. A whole complex of issues arises from our discussion of the Temple of Lemminkäinen and its treasure, a labyrinth that – coincidentally – is crystallized in a piece of art that the American-Japanese artist Ken Hiratsuka created at the site in the early 1990s. Although formally not a labyrinth design, the spiral-dominated image pecked in the rock has affinities with the labyrinth motif, which in turn brings together a number of broader themes related to the relationship between the worlds that are overground and underground.

Ior Bock, the Temple of Lemminkäinen and the Bock Family Saga

The eccentric mystic Ior Bock (original name Bror Holger Svedlin, 1942–2010) was a well-known mythologist and fringe character in Finland in the 1980s and 1990s (Ior Bock, n.d.). His pseudonym 'Ior' was taken from the character Eeyore in *Winnie the Pooh*, whose melancholy moods were shared by the young Bock, who spent some of his childhood in orphanages. At the age of twenty he was involved in a mysterious incident in which his brother Erik was shot and killed, either accidentally by Bock as he initially claimed, or – as he stated later – in a suicide (Londen, 2010; Hay, 2022). In the late 1960s, Bock gained fame as a tour guide in Helsinki, but he was made redundant in 1984 because of his increasingly fanciful stories. Bock had been frequenting the hippie circles of Goa since the mid-1970s and started to attract followers among yogis and spiritualists in Finland and Goa (Londen, 2010). Bock was also relatively open about using cannabis, which was prohibited and highly suspicious in Finland in the 1980s. Bock was widely known in Finland – something of an offbeat celebrity – but he and his followers operated in the margins and underground of Finnish society. Bock was eventually stabbed to death by one of his followers/caregivers in 2010.

After his mother's death in 1984, Bock started telling an alleged family story, styled as the Bock Family Saga, which according to him had run in the family since time immemorial (Borgen, 2018). Bock and his hippie community settled at Bock's family estate in Gumbostrand in Sipoo, southern Finland, about 30 kilometres east of Helsinki. The community developed an alternative history of the Bock family and the entire humankind, and anchored it in certain places in Finland. According to Bock, a great treasure was hidden and sealed under the rock outcrop, which Bock dubbed as the Temple of Lemminkäinen (Figure 6.1), in AD 987 before the fictitious invasion of Christian crusaders into Finland.

According to Borgen (2018: 170), 'The Bock family's treasure chambers are a conically shaped underground cavity which was created by centrifugal forces around the North Pole during the formation of the planet.' At this point in time, according to Bock, the North Pole was on the outskirts of what is now Helsinki. Bock claimed that five million generations each added a room to the treasure chamber, cut into the rock: the resulting tunnel spiralled down some 10,000 kilometres, filled with treasure and untouched since it was sealed shut in AD 987. Borgen (2018: 171) states rather optimistically that 'When the storehouse is finally opened, all truth about these speculations will hopefully be revealed.'

Ior Bock told his followers that the temple would open and become accessible in 1987, on the 1,000th anniversary of its sealing, but when nothing happened, they decided to excavate and open the cave themselves. The work commenced in 1987 and was briefly supported by a large Finnish construction company, Lemminkäinen Group Ltd., although the company retreated when it became known that drugs were used in Bock's commune (Londen, 2010). Lemminkäinen is one of the main heroes of the Finnish national epic, the *Kalevala* (Lönnrot, 1835), based on Karelian folk poetry but heavily edited and modified by the compiler/author of the work, Elias Lönnrot. Bock claimed that his family saga reflected original Finnish pagan knowledge and tradition, and his story was published in 1996 as a version compiled and edited by his followers under the title *The Bock Family Saga: The Mythology of Väinämöinen* (our translation; Javanainen, 1996). As Väinämöinen is the central character of the *Kalevala*, Bock effectively styled himself as the inheritor and flame-bearer of ancient mythical heroes and traditions that he traced back thousands of years before the last Ice Age and ultimately the beginnings of humankind.

Starting in 1987, Bock's followers cut into the rock with pickaxes, shovels, stone drills, explosives and excavation machines. Based on Bock's description, and creative interpretations of Ground Penetrating Radar surveys, they expected to break through into a tunnel leading to a large 'waterlock' room: instead, they found a solid wall of rock. In their writings on the excavation work in progress, Bock's followers express their frustration at the lack of support from sponsors, and from Finnish museums or heritage agencies. They turned instead to private, international sponsorship, offering exclusive rights to the photographs of the treasure chamber upon its discovery (Positive Foundation, 1992). Excavations continued into the early 1990s until money started running out.

As Bock was the landowner and the site is not officially recognized as archaeological heritage, the group had free hands to work as they pleased, but nothing was found. In 1987, an archaeologist from the National Board of Antiquities (now the National Heritage Agency) inspected the site, which he deemed of no archaeological interest, but he noted that 'the hippies had dug [. . .] with a digging machine to over 4 metres deep' (Seger, 1987). Bock's group organized ground-penetrating radar surveys at sites connected to the Bock Saga, producing various scientific-looking diagrams and maps of the alleged structure of the temple and other buildings. By the late 1990s, Ior Bock had accumulated a great deal of debt, largely because of the excavations, and his family estate was auctioned to cover them. A Helsinki-based yoga school, run by some of Ior's followers, stepped up and bought the temple plot because 'the Temple of Lemminkäinen represents spiritual capital and Finnish base(rock) values, but this spiritual level goes far beyond the understanding of some others' (Londen, 2010, our translation). The excavations were nonetheless put on hold soon after the auction in 1998.

Digging recommenced in 2021 by an international team called the 'Temple Twelve', with the stated aim of uncovering the treasure. We visited the site when the work was underway and interviewed one member of the team. Only two of the original excavation team remain, as many have passed away over the years and much of what was told to us about the temple was second-hand information. This included the story related to the petroglyph on top of the rock above the temple (Figure 6.2). Our interviewee told us that the current team regard the petroglyph as an ancient spiritual and treasure map illustrating the spiral passageway leading into the treasure chamber in the bedrock. However, we identified the carving as a work of Ken Hiratsuka, having just recently engaged with his other work in Kilpisjärvi, Lapland, and consequently familiarized ourselves with his rock carvings around the world (Hiratsuka, n.d.). Other sources suggested that the petroglyph at the temple is indeed of a recent origin (Koivisto, 2007) and, eventually, Hiratsuka himself confirmed that it is his work from the early 1990s. Hiratsuka's piece nonetheless served as a spiritual and literal treasure map for the Temple Twelve and came to serve as a kind of a map for us as well, as we explored the interconnections between a host of issues and themes related to treasures, extractive practices and the extraordinary subterranean realm.

Figure 6.2 Ken Hiratsuka's rock carving above the temple, highlighted with snow (Photo: Oula Seitsonen, 2021).

Spirals and labyrinths

The story recounted in this chapter – or our journey into a labyrinth with the Temple of Lemminkäinen at its centre – began in the very opposite end of Finland from Sipoo. We were doing fieldwork in Kilpisjärvi in the northwesternmost corner of Finnish Lapland, part of the ancestral homelands of the indigenous Sámi. Several landscape features, particularly the imposing fjells in Kilpisjärvi, have been 'painted' with Sámi-inspired mythological narratives of ancient times.

The Kilpisjärvi landscapes were, of course, meaningful to the Sámi who inhabited them long before Finnish settlers in the region, but the most well-known mythical narratives associated with the fjells are Finnish inventions or appropriations of Sámi mythologies of the twentieth century towards developing tourism (Valtonen, 2019), which resonates with the invented mythology of the Temple of Lemminkäinen and the Bock Family Saga. Over the past few years, we have climbed up one of these landmark fjells (Saana) in Kilpisjärvi several times and wondered about the large stone spirals constructed just below the peak, which comprise a natural 'platform' for admiring the surrounding landscape (Kilpisjärvi Science Trails, 2023). We took it for granted that they were new constructions, but it was only during our 2020 fieldwork when we started pondering who had constructed them, when and why?

It was during the same trip in 2020 that one of our students discovered a rock carving on the shore of Lake Kilpisjärvi. This comprised a single pecked line dominated by a spiral and other less easily identifiable motifs. We had barely started speculating on its age and meaning when a local acquaintance informed us that it was made by Ken Hiratsuka in the early 1990s. This find prompted some reflections on spirals and labyrinths (Herva, 2021). When the search for the Lemminkäinen treasure was about to restart in autumn 2021, one of us visited the site only to discover another piece of Hiratsuka's work there (it was known that there is a 'carving' at the site, but until then the maker was unknown). Apparently, Hiratsuka had spent some time there on his trip to Finland in the 1990s when he had been invited to take part in a shaman festival (it turned out that he had also made some pieces in a few other locations in Finland during his visit). The treasure hunter group knew the carving but apparently not its origins, or else chose to regard it as an ancient 'map' to the treasure chamber, which according to Ior Bock (Javanainen, 1996: 48) is reached through a 10,000 kilometre downward expanding spiralling passageway. It is the entrance to this spiralling passageway that the diggers have tried to reach since the 1980s at the temple.

The spiral, circle and labyrinth are different designs but are related to each other visually and cultural-historically. The spiral and labyrinth both imply circular movement going deeper or 'inwards'. Carvings of spirals and concentric circles are known since the Neolithic and Bronze Age, and both designs have associations with the labyrinth motif which is, in a sense, a more elaborate version of the spiral and concentric rings. Spirals have been carved in, for instance, megalithic tombs, which in turn have subterranean and cave-like qualities to them. Carvings of concentric rings may have links to the classical 'Cretan' labyrinth design, and both labyrinth and spiral have sometimes been interpreted as symbolic or spiritual maps (see McCullough, 2005: 11). Furthermore, there are petroglyphs in North West Spain associated with tin mining, which 'might have functioned as a sort of guide ... as a symbol of hope that miners would return to the earth's surface, and, at the same time, as a magical assurance thereof' (Bord, 1976: 10, quoted in Falahat, 2013: 54), which has a resonance with the treasure hunters' interpretation of Hiratsuka's carving at the temple.

Within medieval to contemporary European folklore, circling (and, by extension, circles) are associated with 'opening up' an entrance to the other- or underworld (Menefee, 1985). These circling rituals have often been performed at archaeological sites, which the Temple of Lemminkäinen was/is believed to be by Bock and his followers (and has *become* through associated narratives and its 'excavation'). The association of circles and spirals with stone structures was also extant in prehistory, as suggested by the presence of these motifs in megalithic tombs, such as at Newgrange (McCullough, 2005: 11). The underworld is elementary to the myth of the labyrinth as well: the idea of the labyrinth was perhaps inspired by the limestone cavern systems on Crete (Kotsonas, 2018). Going into the labyrinth in the classical myth is associated with a journey into the underworld which, as already indicated earlier, is an otherworldly and extraordinary domain readily symbolized by the monstrous minotaur. The very act of journeying is important as well: not only does it emphasize that the very *doing* – moving in the labyrinth – is central to its meaning, but also that journeying into the centre of the labyrinth is a rite of passage, which again links the labyrinth to the otherworldly underworld and (personal) transformation (Brunon, 2010).

Hiratsuka's rock carving at the Temple of Lemminkäinen resonates with a long tradition of the labyrinth, spiral and circle/circling. There is also a prominent labyrinth-making tradition in Finland and Fennoscandia, particularly on the Gulf of Bothnia and the coast of the Arctic Ocean and the White Sea. Some are associated with Sámi burial grounds (Olsen, 1991), others with the monasteries on the White Sea, but the functions and meanings of many are unknown, although they are often located on the coastal loci and thus the boundary between the cosmologically significant border between land and sea (e.g. Westerdahl, 2005; Herva and Lahelma, 2020: 104–7). This resonates readily with the traditional shamanistic cosmology of the Finns, which is reflected in the *Kalevala* and the folk poetry that inspired it. The (shamanistic) underworld was accessed, for instance, through clefts in the rock and that realm thus existed, in some sense, under the surface of rock, ground and/or ordinary everyday reality (e.g. Siikala, 1992).

All these diverse associations of spirals, circles and labyrinths offer insights into the treasure hunt at the temple and their wider entanglements, sometimes direct and sometimes indirect, with broader perceptions of the subterranean in various contexts, including archaeology and contemporary extractive industries (Figure 6.3). In what follows, we attempt to unpack the complex relationships between caves, the cultural ideas of the subterranean, accessing the underground realm, engaging with and journeying in the underworld, and searching for riches and treasures. The notion and idea of the temple is also instructive here, as it

Figure 6.3 A spiral drawn in the snow by someone above the entrance to the Temple of Lemminkäinen (Photo: Oula Seitsonen, 2021).

readily echoes with matters of the spiritual and otherworldly, as well as with underground treasures that also have spiritual dimensions both in the past and present.

Discovering unknown pasts and places

The Temple of Lemminkäinen is not an isolated flight of fancy, but is a part of the wider Bock Family Saga of ancient secrets. Ior Bock was a charismatic storyteller and it was notoriously difficult to tell how serious he was with his family saga of ancient traditions (published as Javanainen, 1996), but one branch of the story was that the last pagan king of Finland sealed a treasure under the Gumbostrand rock outcrop, while other valuables, such as gold objects and precious stones, were hidden elsewhere. Objects of particular significance, according to the saga, included three golden ram statues, two of which had been hidden at an old sacrificial forest in Snappertuna, South West Finland, and one at the castle of Kajaani in North East Finland. This castle was built in the Swedish-Russian borderland in the seventeenth century, when Finland was part of Sweden, but the Bock Saga holds that there was a much older castle at the site that had served as a hiding place for pagan kings during the Swedish crusades to Finland (Javanainen, 1996). The story of the hidden ram statue surfaced in the media in the 1990s when a ground-penetrating radar survey in the castle yard suggested that a sizable metal object was buried deep in the courtyard. Subsequent archaeological excavations in the early 2000s revealed a coil of cable that archaeologists determined to be the most likely source of the radar-signal (Rakennusperintö.fi, n.d.).

This did not discourage the search for the treasure at the Temple of Lemminkäinen. Bock did not require that the diggers believe in the treasure: he himself said that he did not necessarily believe in the material existence of the treasure. If nothing substantial turned up, it would just show that the story of the treasure is a narrative device for something that is of a symbolic and spiritual nature. Stories of hidden treasures – however implausible they may seem – have a captivating power, as has repeatedly been seen in different contexts. It is always possible that there might be *something* behind such fabulations, which in turn resonates with the mysterious character of the subterranean, as discussed later. The Bock Saga perhaps makes little sense in light of the 'known' and established narratives of the past – but it is also true that there is a long history of extraordinary discoveries underground ranging from treasures to lost civilizations, with the Minoans and Mycenaeans a case in point. Why should history have come to an end in this respect?

In order to put the Bock Saga in a broader context, it can be usefully compared to early modern antiquarian pursuits for discovering unknown pasts, which, like the Bock Saga, may appear strange from today's perspective. The Bock Saga draws from Finnish and Scandinavian mythologies and numerous other sources, mixing them in an eclectic manner that resembles the way that early modern Swedish antiquarians sought to reconstruct the past and uncover hitherto unknown aspects of it. We can take, for example, the work of Olof Rudbeck's key idea that Atlantis had been in Sweden, and Atlantis – and by extension Sweden – was the original homeland of civilization (Anttila, 2014). Rudbeck was not alone with his alternative-historical pursuits in early modern Sweden and Europe. Olaus Magnus, for instance, had reconstructed similar histories of Sweden in the sixteenth century, whereas Johannes Messenius chronicled – while imprisoned in the castle of Kajaani – a fabulous history of his country and nation in the seventeenth century, which has a direct thematic link to Ior Bock's narrative. This so-called Gothicist historiography fell out of vogue with the collapse of the Swedish Empire in the early-eighteenth century, but the 'mythical' historiography did not die away. Besides the Bock Saga, the *Kalevala* and the folk poetry behind it have inspired generations of artists and scholars since the nineteenth century. The results were contentious at best, but the point here is that there was a continued appeal to discovering something 'real', in a historical sense, under the surface of folk poetry.

And why not – in principle, anyway? Some of the greatest discoveries that have shaped European mindscapes in the nineteenth and twentieth centuries revolve around the discovery of the material realities behind the Homeric epics, such as Troy and Mycenae, as well as the 'palace of Minos' at Knossos, recounted in classical Greek mythology. Heinrich Schliemann, Arthur Evans and other pioneering researchers of the Aegean Bronze Age effectively revealed a historical reality behind myths that, in a mainstream view at the time, were just that – myths and narratives without any real connection to past realities. Likewise, more recently, Andrea Carandini (2011) and his team claim to have discovered some historical reality behind the origins, myths and mythical histories of early Rome (although they do not argue that myths are literally true in the same way as Schliemann did).

Closer to the case of the Bock Saga, the historian Kyösti Julku (1986) tackled the question of 'Kvenland' in the 1980s. Kvenland is a semi-mythical realm mentioned in Norse sagas and other (later) sources, which has puzzled scholars for a long time. In some Norse accounts, Kvenland is said to have been ruled by a king called Faravid, which resonates with the contemporary pseudohistorical fascination with forgotten Finnish pagan kings and kingdoms. Julku did not argue for the existence of a Kven kingdom or kings, but he argued that, perhaps, there once existed a power in northern Fennoscandia that left 'echoes' in later sources. His theory remained obscure for many years as the prevailing historical narratives held that the Swedes and the Rus struggled to control the northernmost reaches of Europe during this period. This was a 'fact' until it was challenged – and along the lines that Julku had suggested. Namely, archaeological research now suggests that there was a 'third power' in operation there that delimited the activities of the Swedes and the Rus in the early modern period. Indeed, this later research has directly and indirectly corroborated the view that a trader network controlled the much-desired northern goods, particularly fur and fish (e.g. Bergman and Edlund, 2016; Kuusela et al., 2018; Nurmi et al., 2020).

Wonders and treasures beneath the surface

The history of antiquarian and archaeological research from the early modern period through to the present shows that digging can reveal spectacular and unexpected traces and dimensions of the past. This potential and allure of the subterranean must also be considered and framed within a yet broader perspective of how worlds under the surface of the ground have been culturally conceived, perceived, experienced and engaged with over centuries and millennia.

Wondrous as well as mundane things have always been lost and found in the ground. Natural processes and human activity – particularly digging, for whatever specific purpose – hint at the secrets and mysteries of that unknown world in the guise of fossils, rock crystals, gold nuggets and ancient objects, among others. Such strange things found underground, together with the peculiar sensory experiences of underground spaces, have long demonstrated the extraordinary otherness of the subterranean realm. This alien subterranean realm has intrigued people from Palaeolithic cave-painters to modern-day miners and urban explorers. The subterranean has been both deeply unfamiliar and critically important to people for thousands of years. People may not be ordinarily (or consciously) aware of how ideas and imageries of the underground world over innumerable generations have shaped their mindscapes, but folklore and popular culture illustrates the cultural ideas associated with the underground: almost anything extraordinary can be found underground, from chests of pirate gold to a time portal in a cave, and from buried Second World War aeroplanes in Myanmar to a Nazi treasure-train under a mountain in Poland (e.g. Ferguson, 2020). Within this cultural scaffolding of the real and imagined undergrounds and what they contain, there is nothing inherently or particularly aberrant to the idea that ancient treasures might just as well be discovered in southern Finland, however unlikely it may seem.

The treasure hidden under the Temple of Lemminkäinen is said to be of unimaginable riches, including golden statues. Even the temple itself is supposed to be coated with gold (Javanainen, 1996: 48). Coincidental or not, golden statues strike a chord with the Norse *Óláfs saga helga* (Óláfs saga helga, 1945; see Vasaru, 2014) that describes a Viking raid, in collaboration with Kvens, to the mythical realm of Bjarmia, where a temple housing a valuable cult statue with gold and silver offerings is looted. Whatever the inspiration and story behind the golden statues in Ior Bock's narrative, gold, of course, is a paradigmatic substance of real and fictitious treasures, whether in the form of the gold objects that Schliemann discovered in the shaft graves of Mycenae, caches of Spanish doubloons or hidden Nazi gold bars. Indeed, public perceptions of archaeology tend to revolve around (magnificent) discoveries and finds (Lucas, 2001: 19).

As 'dreams materialized' (D'Angelo and Pijpers, 2018: 219), gold is a substance that can infect people with 'gold fever' and drive them literally or figuratively insane, as illustrated by the nineteenth-century gold rushes around the world, including Finnish Lapland. But gold is just one example of the wondrous things and riches of the subterranean world that, upon retrieval, affect the world above ground. Metals are a source and repository of wealth, but have also been agents of socio-cultural transformations, especially since the Bronze Age. The extraction, production and uses of metals and other subterranean substances, such as coal and oil, are usually described in practical and rationalist terms. But they are products of the mysterious subterranean, which is reflected in historical folklore as well as contemporary perceptions of and attitudes to extractive industries (see below).

In early modern times, orebodies – like treasures – were explicitly associated in folk and learned views with their guardian spirits that ultimately controlled and manipulated them. Engagement in mining and treasure hunting thus involved interacting with supernatural beings and powers of the subterranean realm (Sarmela, 1994; Fors, 2006; Dillinger, 2011). There is, moreover, ample folklore about the supernatural dimensions of the underground and its inhabitants, with the general aura of the magical and extraordinary more or less explicitly attached to the different products of the underground. Both orebodies of valuable minerals and treasures have been considered volatile and erratic because of the non-human powers governing them (Sarmela, 1994; Fors, 2006; Dillinger, 2011), not dissimilarly from how mining projects today are governed by mysterious and unpredictable forces, especially the 'invisible hand of capital' (see Moeran and de Waal Malefyt, 2018 on capitalism as a form of magical thinking).

Treasures and mining both revolve around dreams of riches and a better future, and are assigned with socially (and spiritually) transformative powers. But securing underground riches often fails to deliver what they promise, and can result in social and environmental disasters in the case of mining projects. Mining projects are represented as calculated pursuits governed by rationality, but under this surface they actually operate similarly as treasure hunting: driven by hopes and dreams that frequently turn out to be unsubstantiated, and abilities and possibilities overestimated. But treasures are *sometimes* discovered just like mines have *sometimes* been successful in generating and distributing wealth. Many ventures may have failed, but perhaps *this one* will be successful (fingers crossed) (Komu, 2019).

It is only fitting, then, that rich orebodies are sometimes referred to as treasures in the media (Komu, 2019; Herva et al., 2022: 168). This could be taken simply as a casual metaphor, but it actually crystallizes the magical thinking operating underneath the rationalist surface of extractive industries as a form of engaging with the strange subterranean world. If centuries of constant failures in the supposedly calculated and rationalist treasure hunting that is industrial mining have not eradicated faith in the practice, why should failures compromise faith in 'true' treasure hunting? Both unfold in relation to the unknown and magical underground, even if the role of this factor encompassing faith, dreaming and magical thinking is usually not directly, explicitly or consciously identified and recognized in the context of contemporary industrial mining.

The treasure hidden under the Temple of Lemminkäinen may appear idiosyncratic and outlandish, but it actually reflects and replicates older and broader cultural ideas about the extraordinary underground and its erratically behaving riches. Whatever wishful thinking the Lemminkäinen treasure – or the idea of the treasure in general – may command, 'rationalist' mining projects (and the wealth they are believed to generate) are similarly steeped in and entangled with all kinds of beyond-the-rational ideas (see Komu, 2019 and below). Moreover, the 'excavation' of the temple more closely resembles a mining or quarrying project than archaeology: digging into solid rock to find a seam of physical (and spiritual) treasure.

Digging as journeying in the underground

The hunt for the Lemminkäinen treasure has lasted for more than thirty-five years. This strikes a chord with historical and folkloric accounts of the difficulties and intricacies involved in locating and retrieving treasures, often interrupted by bad weather or other (super)natural phenomena. Finding treasures and digging them up has traditionally required occult knowledge and procedures because treasures (e.g. Dillinger, 2011), as discussed earlier, often have supernatural qualities and guardians. Similarly, mining requires finding orebodies suitable for industrial exploitation and extracting mineral riches requires huge investments of time, knowledge and other resources including technology and labour.

The location of the Lemminkäinen treasure is 'known' but retrieving it should perhaps not be regarded as a simple mechanical task, but as imbued with meaning in itself, a mythical and transformative journey through solid rock. A comparison can be made to shamanistic journeying in the underworld or journeying through the labyrinth. In fact, the Bock Saga refers to the underground tunnel system beneath the Gumbostrand rock outcrop as the 'Undervärlden'

(Underworld) (Borgen, 2018), and journeying through corridors and passageways is integral to the functionality of the labyrinth as an otherworldly environment. In this view, Hiratsuka's petroglyph at the temple is a fitting treasure map: finding the treasure is not only about the location ('X marks the spot'), but about the journey, which involves encountering and engaging with an underground world literally, figuratively and metaphysically. Indeed, the idea of the treasure and the journey to it may indeed be more important than the actual treasure.

The digging for the Lemminkäinen treasure journeys underground through rock, which echoes shamanistic travelling and is associated with, among other things, cracks, clefts and openings in the surface of rocks (e.g. Herva and Lahelma, 2020). This brings a spiritual dimension. In folk poetry – which has been one inspiration for the Bock Saga – going underground is associated with the search for wisdom. For example, when Väinämöinen was making his (magical) boat, he went on a journey inside Antero Vipunen, an ancient giant and sage who was buried and/or grown over, to gain certain magic words, while Odin gained his wisdom by hanging himself and thus venturing into the realm of the dead, which provided him with special knowledge (e.g. Siikala, 2012).

Similarly, the very act of excavating the temple echoes ostensive practice in folklore. Ostension is the term used within folkloristics to describe the 'acting out' of a legend in some way (see Dégh and Vázsonyi, 1983). This is not necessarily a re-enacting of the contents of the legend, but behaving in ways that support the narrative. The excavation here is thus a kind of ostensive act in relation to the Bock Saga. By engaging in digging and searching for the alleged treasure, the participants are effectively taking part in the mythical narrative that Ior Bock crafted, and hence contributing to making the Bock Saga real through enacting the story. This compares to early modern enactment works as a way of invoking past (mythical) worlds, connecting to and engaging with a living past, which involved historical-mythical performances, but also excavating (Godwin, 2002; Herva and Nordin, 2015). In this view, it is the doing that matters, the pursuit itself, and not (only) discovering and securing a fabulous treasure. Similar ideas and practices are also present in Lapland gold prospecting, past and present.

Whether or not these myths have any direct bearing on the hunt for treasure under the Temple of Lemminkäinen, the pursuit can be seen as having broadly spiritual dimensions to it. As Ior Bock himself admitted, he was not sure if the treasure is something tangible with a monetary value to begin with; it might also be something metaphysical and spiritual. As to the spiritual dimension of digging, it is also relevant that yoga was an important element of Ior Bock's life and that of his circle, and indeed the site is now owned by a yoga school of his followers. Thus, there is an element of personal transformation and self-realization to the broader setting of the search for the treasure. Indeed, 'What most captured Carl and his friends was the promise from Bock that a second Paradise Time could arise if only the priceless artefacts of the Aser could be uncovered from the frozen wastes in which they were still buried' (Ayaga, 2021), which is essentially the spiritual dimension of the promised treasure.

Somewhat less obviously – but critically importantly – historical and contemporary industrial mining in Finland is also not only about financial profit, but mining has been and is invested with meaning in itself. The first industrial mining boom in northern Fennoscandia took place in the seventeenth century and was fostered by dreams of colonial wealth that a discovery of silver ore sparked off (Nordin, 2020: 116). The dreams did not come true, but the state of Sweden continued to support mining endeavours not only for economic profit, but also as a form of social engineering: industrial work was a means of 'disciplining' and 'bettering' the northerners living on the margins of state control (Naum, 2018).

Contemporary mining projects and other megaprojects can be surrounded by messianic expectations of 'improvement', especially of socio-economically challenged regions (cf. Harvey and Knox, 2012), as exemplified by the story of the Talvivaara mine in Finland (e.g. Sölbar, 2021). Megaprojects incur formidable sunk costs to feed dreams and fantasies, which – like pseudoarchaeological pursuits – are easy to dismiss as individual irrationality or foolishness but are embedded in broader socio-cultural imaginaries. Finland is very open to mining projects, but the expected benefits seem to be only partially realized. That is, extractive industries do create jobs but those are temporary and not nearly on the same scale as in the past, because of widespread automation (Paredes and Fleming-Muñoz, 2021; see also Geenen and Gleiberman, 2023). Also, the state does not generate any significant income from mining but effectively gives mining rights and mineral resources for free to (often foreign) companies with continued political support to do so. This is also mirrored in a recent Finnish newspaper comic related to the selling of the Temple of Lemminkäinen abroad (Leppänen, 2022), sarcastically commenting that 'well, now the treasure will definitely be found there' (Vainio, 2023).

One way of making sense of this peculiar situation is that magical thinking is at play, albeit of course not identified or regarded as such. Mining is considered to bring about wealth and wellbeing as if by magic, in that the mechanisms of this remain unclear given the actual operational environment of extractive industries in Finland (see above). It seems that nostalgic or 'retrospective' dreams and fantasies are in operation here; mining futures are imaginarily modelled after past experiences. In post-Second World War decades, mining and other industries generated national wealth in Finland and transformed a formerly poor and underdeveloped country into a Nordic welfare state. This is also why industries have been central to Finnish identity and nation-building throughout the twentieth century, and industries consequently have a magical aura to them. However, extractive industries and their operational and financial environment have changed radically from those times and what worked then simply does not work now. The idea of generating and distributing wealth through extraction is only partly founded on realities, and is largely driven by nostalgic dreams and memories (see also Sölbar, 2021). And therein lies a further connection to the treasure in the Temple of Lemminkäinen: dreaming of lost utopian worlds and trying to bring them into being by accessing treasures from underground.

Conclusion: The lure of underground

Many would be reluctant to label such pseudoarchaeological fantasies as the Bock Saga and Temple of Lemminkäinen as 'heritage', but we have seen that the narrative and its ostensive enactment through the excavation of the 'temple' are closely entangled with heritage matters which, moreover, have also direct and indirect links to extractive industries as regards ideas of the subterranean realm as a deeply ambiguous domain and a place of origin of metals and mineral riches. We have explored this intangible heritage of extractive industries from the perspective of the temple, towards putting cultural ideas and perceptions of the strange underground and the search and extraction of riches from below ground on a sharper relief. The *specific* forms and expressions of the ideas and imaginaries of underground are obviously somewhat different in the case of pseudoarchaeology from industrial extraction, but essentially similar perceptions of the subterranean feature in both fields. It will be elementary, as regards developing more sustainable mining, to become better aware of how 'magical thinking' operates in various practices of resource extraction, albeit often implicitly and indirectly.

Treasures are capricious and awe-inspiring things of an uncertain nature. This is probably one reason why they fascinate people in the first place and are, in that respect, similar to the deeply ambivalent subterranean domain as a world of illusions, misperceptions, transformations, danger, liminality and uncertainty – places where things may appear to be real but are actually not or may appear to be one thing but then turn out to be something else. The search for the treasure in the Temple of Lemminkäinen echoes various cultural imaginaries of the subterranean world. This is partly because it involves a cave and partly because it involves digging, along with the more general idea that strange and valuable things can be discovered below ground, in this particular case the prospect of discovering the greatest treasure of all time, whether in the guise of material wealth and/or spiritual transformation in the heart of the figurative and alleged labyrinth that is the temple and symbolized by Hiratsuka's rock carving. Indeed, Ior Bock (n.d.) himself said in a newspaper interview that he does not necessarily believe in the temple as a real-world place, but that part of the saga may just as well be symbolic and offer spiritual guidance. Thus, although the Bock Family Saga may initially strike as an obscure and eccentric work, it affords perspectives, as we have discussed in this chapter, on various broader and more 'serious' or mainstream questions of how the real and culturally imagined subterranean realm is entangled within historical and contemporary societies.

Acknowledgements

This research is part of the project 'Extractive Industries as Engagement with the Extraordinary Subterranean: Culture, Heritage and Impact of Resource Extraction in Northernmost Europe', funded by the Academy of Finland (Research Council of Finland), decision no. 339483.

References

Anttila, T. (2014), *The Power of Antiquity: The Hyperborean Research Tradition in Early Modern Swedish Research on National Antiquity*, Oulu: University of Oulu.

Ayago, K. (2021), '"Temporarily Insane" by Carl Borgen', *The London Economic*, 1 November 2021, available online: https://www.thelondoneconomic.com/entertainment/temporarily-insane-by-carl-borgen-298990/ (accessed 1 April 2022).

Bergman, I. and L.-E. Edlund (2016), '*Birkarlar* and Sámi – Inter-Cultural Contacts Beyond State Control: Reconsidering the Standing of External Tradesmen (*Birkarlar*) in Medieval Sámi Societies', *Acta Borealia*, 33 (1): 52–80. doi:10.1080/08003831.2016.1154676.

Bock Saga Official (n.d.), 'Home. Bocksaga. Official', available online: www.bocksaga.info/ (accessed 10 February 2021).

Bord, J. (1976), *Mazes and Labyrinths of the World*, London: The Anchor Press.

Borgen, C. (2018), *The Bock Saga: An Introduction*, Self-published.

Borgen, C. (2021), *Temporarily Insane*, Self-published.

Brunon, H. (2010), 'Labyrinth', in A. Grafton, G.W. Most and S. Settis (eds), *The Classical Tradition*, 505–6, Cambridge MA: Belknap.

Bywater, T. (2021), 'Finnish Treasure Hunters Reportedly Close to Finding Ior Bock Fortune', *New Zealand Herald*, 4 November 2021, available online https://www.nzherald.co.nz/travel/finnish-treasure-hunters-reportedly-close-to-finding-ior-bock-fortune/GIHZ33TZB4FGIAUJXPWXAH5V34/ (accessed 1 April 2022).

Carandini, A. (2011), *Rome Day One*, Princeton: Princeton University Press.

D'Angelo, L. and R. J. Pijpers (2018), 'Mining Temporalities: An Overview', *The Extractive Industries and Society*, 5 (2): 215–22. doi.org/10.1016/j.exis.2018.02.005.

Dégh, L. and A. Vázsonyi (1983), 'Does the Word "Dog" Bite? Ostensive Action: A Means of Legend-Telling', *Journal of Folklore Research*, 20 (1): 5–34.

Derricourt, R. (2012), 'Pseudoarchaeology: The Concept and Its Limitations', *Antiquity*, 86 (332): 524–31. doi:10.1017/S0003598X00062918.

Dillinger, J. (2011), *Magical Treasure Hunting in Europe and North America: A History*, Basingstoke: Palgrave.

Fagan, G. G. (ed.) (2006), *Archaeological Fantasies: How Pseudoarchaeology Misrepresents the Past and Misleads the Public*, London: Routledge.

Falahat, S. (2013), *Re-imaging the City: A New Conceptualisation of the Urban Logic of the 'Islamic City'*, New York: Springer.

Ferguson, J. (2020), 'Spitfires Sprouting in the Burmese Spring: The Real-life Quest for Historic Fantasy Aircraft in Contemporary Myanmar', *TRaNS: Trans-Regional and National Studies of Southeast Asia*, 8 (2): 135–46.

Fors, H. (2006), *The Limits of Matter: Chemistry, Mining, and Enlightenment*, London: The University of Chicago Press.

Geenen, S. and M. Gleiberman (2023), 'Superfluous Jobs in Extractive Industries: The Usefulness/Uselessness of Job Creation after Dispossession', *Work, Employment and Society*, 37 (2): 394–411. https://doi.org/10.1177/09500170211008721.

Godwin, J. (2002), *The Pagan Dream of the Renaissance*, Grand Rapids: Phanes.

Harvey, P. and H. Knox (2012), 'The Enchantments of Infrastructure', *Mobilities*, 7 (4): 521–36.

Hay, M. (2022), The Strange Story of Ior Bock, *InsideHook*. https://www.insidehook.com/article/history/strange-story-ior-bock.

Herva, V.-P. (2021), 'Minoan Lapland: Fieldwork, Spirituality and Connecting across Time and Space', *Time and Mind*, 14 (3): 397–415.

Herva, V.-P. and A. Lahelma (2020), *Northern Archaeology and Cosmology: A Relational View*, Abingdon: Routledge.

Herva, V.-P. and J. M. Nordin (2015), 'Unearthing Atlantis and Performing the Past: Ancient Things, Alternative Histories and the Present Past in the Baroque World', *Journal of Social Archaeology*, 15 (1): 116–35.

Herva, V.-P., T. Komu and T. Paphitis (2022), 'Extraordinary Underground: Fear, Fantasy, and Future Extraction', in S. Sörlin (ed.), *Resource Extraction and Arctic Communities: The New Extractivist Paradigm*, 166–82, Cambridge: Cambridge University Press.

Hiratsuka, K. (n.d.), 'Home'. Available online: www.kenrock.com (accessed 24 December 2021).

Hunt, W. (2019), *Underground: A Human History of the Worlds beneath Our Feet*, London: Simon & Schuster.

Ior Bock (n.d.), 'Ior Bock – virallinen sivusto'. Available online: www.iorbock.com (accessed 6 January 2021).

Javanainen, J. (ed.) (1996), '*Bockin perheen saaga: Väinämöisen mytologia. Ior Bock*,' Helsinki: Synchronicity.

Julku, K. (1986), *Kvenland–Kainuunmaa*, Rovaniemi: Pohjois-Suomen Historiallinen Yhdistys.

Kilpisjärvi Science Trails (2023), 'Saana Labyrinths', available online: https://kilpisjarvisciencetrails.stqry.app/1/tour/12180/item/155484?show-detail-modal=true (accessed 14 June 2023).

Koivisto, S. (2007), 'Sipoo. Manneralueen esihistoriallisen ajan ja saariston esihistoriallisen sekä historiallisen ajan muinaisjäännösten inventoinnit', unpublished survey report, Helsinki: National Heritage Agency.

Komu, T. (2019), 'Dreams of Treasures and Dreams of Wilderness: Engaging with the Beyond-the-Rational in Extractive Industries in Northern Fennoscandia', *The Polar Journal*, 9 (1): 113–32. https://doi.org/10.1080/2154896X.2019.1618556.

Komu, T. (2020), *Pursuing the Good Life in the North: Examining the Coexistence of Reindeer Herding, Extractive Industries and Nature-Based Tourism in Northern Fennoscandia*, Oulu: University of Oulu.

Kotsonas, A. (2018), 'A Cultural History of the Cretan Labyrinth: Monument and Memory from Prehistory to the Present', *American Journal of Archaeology*, 122 (3): 367–96.

Kroonenberg, S. (2013), *Why Hell Stinks of Sulfur: Mythology and Geology of the Underworld*, London: Reaktion.

Kuusela, J.-M., R. Nurmi and V. Hakamäki (2018), 'Unhierarchical and Hierarchical Core-Periphery Relations: North Fennoscandian Trade Network from the Middle Ages to the Post-Sixteenth Century', *American Anthropologist*, 120 (4): 765–80. doi: 10.1111/aman.13104.

Leppänen, M. (2022), 'Mystikko lupasi, että Sipoossa on valtava aarre – nyt alue myydään huutokaupassa ja kiinnostusta on ulkomailla asti', *YLE*, 12 November 2022. Available online: https://yle.fi/a/74-20004246 (accessed 31 December 2022).

Londen, M. (2010), 'Holgerin saaga – alias Ior Bockin tarina'. Available online: https://www.magnuslonden.net/fi/artikkelit/haastattelut/article-64846-28436-holgerin-saaga-alias-ior-bockin-tarina (accessed 1 April 2021).

Lönnrot, E. (1835), *Kalevala*, Helsinki: Frenckell ja Poika.

Lucas, G. (2001), *Critical Approaches to Fieldwork: Contemporary and Historical Archaeological Practice*, London: Routledge.

McCullough, D. W. (2005), *The Unending Mystery: A Journey Through Labyrinths and Mazes*, New York: Anchor Books.

Menefee, S. P. (1985), 'Circling as an Entrance to the Otherworld', *Folklore*, 96 (1): 3–20.

Moeran, B. and T. de Waal Malefyt (eds) (2018), *Magical Capitalism: Enchantment, Spells, and Occult Practices in Contemporary Economies*, New York: Palgrave.

Naum, M. (2018), 'The Pursuit of Metals and the Ideology of Improvement in Early Modern Sápmi, Sweden', *Journal of Social History*, 51 (4): 784–807. https://doi.org/10.1093/jsh/shx011.

Nordin, J. M. (2015), 'Metals of Metabolism: The Construction of Industrial Space and the Commodification of Early Modern Sápmi', in M. P. Leone, M. Jocely and E. Knauf (eds), *Historical Archaeologies of Capitalism*, 249–72, New York: Springer. doi:10.1007/978-3-319-12760-6_11.

Nordin, J. M. (2020), *The Scandinavian Early Modern World: A Global Historical Archaeology*, Abingdon: Routledge.

Nurmi, R., J.-M. Kuusela and V. Hakamäki (2020), 'Swedenization of the North – The Early Medieval Swedish Northern Expansion and the Emergence of the Birkarls', *Acta Borealia*, 37 (1–2): 1–26.

Óláfs saga helga (1945), '"Óláfs saga ins helga"'. *Íslenzk fornrit* XXVII. bindi. Heimskringla II. Snorri Sturluson, Heimskringla, 2–451. Bjarni Aðalbjarnarson gaf út. Reykjavik: Hið Íslenzka Fornritafélag.

Olsen, B. (1991), 'Material Metaphors and Historical Practice: A Structural Analysis of Stone Labyrinths in Coastal Finnmark', *Fennoscandia Archaeologica*, 8: 51–8.

Paredes, D. and D. Fleming-Muñoz (2021), 'Automation and Robotics in Mining: Jobs, Income and Inequality Implications', *The Extractive Industries and Society*, 8 (1): 189–93.

Partanen, S. J. (1999), *Sankareita, veijareita ja huijareita: Lapin kullankaivajien tarina*, Tampere: Edita.

Positive Foundation (1992), *Bock Saga Documentation: English/German version*, Berlin: Positive Foundation.

Rakennusperintö.fi (n.d.), 'Uutiset 2006. Kultaisen pukin sijaan löytyi sähkökaapeleita', available online: www.rakennusperinto.fi/news/Uutiset_2006/fi_FI/kultaisenpukin/ (accessed 16 October 2020).

Sarmela, M. (1994), *Suomen perinneatlas: Suomen kansankulttuurin kartasto* 2, Helsinki: Suomalaisen Kirjallisuuden Seura.

Schadla-Hall, T. (2004), 'The Comforts of Unreason: The Importance and Relevance of Alternative Archaeology', in N. Merriman (ed.), *Public Archaeology*, 269–85, London: Routledge.

Seger, T. (1987), '"Lemminkäisen temppeli." Esihistoriallisen kohteen tarkastus. 6.8.1987. Esihist. tsto 162/13.10.1987', unpublished inspection report, National Heritage Agency, Helsinki.

Siikala, A.-L. (1992), *Suomalainen šamanismi: mielikuvien historiaa*, Helsinki: Finnish Literature Society.

Siikala, A.-L. (2012), *Itämerensuomalaisten mytologia*, Helsinki: Finnish Literature Society.

Sölbar, T. (2021), 'The Belief in Mining: How Imageries of Other Mines May Brighten Arctic Minescapes', *Polar Record,* 57: E44. doi:10.1017/S0032247421000188.

The Temple Twelve (n.d.), 'TempleTwelve', available online: https://twitter.com/TempleTwelve (accessed 17 July 2020).

Vainio (2023), 'Pilapiirros / Vainio', *Vantaan Sanomat*, 18 March 2023. Available online: https://www.vantaansanomat.fi/paakirjoitus-mielipide/5799515 (accessed 14 June 2023).

Valtonen, T. (2019), 'Miten Saanasta tuli pyhä? Erilaisten rinnakkaisten Saana-diskurssien tarkastelua', Terra, 131 (4): 209–222.

Vasaru, M. (2014), *Bjarmaland*, Oulu: University of Oulu.

Westerdahl, C. (2005), 'Seal on Land, Elk at Sea: Notes on and Applications of the Ritual Landscape at the Seaboard', *International Journal of Nautical Archaeology,* 34 (1): 2–23.

Wilson, J. A. P. (2012), 'The Cave Who Never Was: Outsider Archaeology and Failed Collaboration in the USA', *Public Archaeology,* 11 (2): 73–95. doi: 10.1179/1465518712Z.0000000007.

7

Uses of the Industrial Mining Past: The Revitalization of Old Mines as New Outdoor Spaces in Sweden

Albina Pashkevich and Daniel Brandt

Introduction

There are contrasting viewpoints regarding the utilization of old mines for outdoor activities. These spaces can indeed offer unique and immersive experiences for visitors. However, effectively conveying the historical significance and cultural heritage of the mining industry to visitors is not as straightforward as it may appear to the general public. Although outdoor activities, such as hiking, biking or kayaking can be facilitated in these areas to attract tourists and promote nature-based tourism, providing visitors with a deep understanding and appreciation of the historical context of the site poses more challenges in interpretation, allowing a better appreciation of cultural and natural values confined to them. Traditionally, these places have been established with the aim of preserving, educating and communicating the cultural heritage of industrial activities (Hospers, 2002; Heldt Cassel and Pashkevich, 2011; Pashkevich, 2017; Bockelmann, 2023). However, at the end of 1990 and the beginning of 2000, the industrial sites became increasingly viewed as arenas providing exciting new opportunities considering their tangible characteristics and making them as subjects for new multifaceted postmodern interpretations (Hospers, 2002; Berger, 2019; Berkenbosch et al., 2022; Bockelmann, 2023). The mining areas of Bergslagen in Sweden marked by centuries-long history of the past and still ongoing mining operations, provide an interesting case study for examining a similar process, particularly in relation to revitalization of industrial spaces initiated in the early 1990s. Furthermore, this chapter also focuses on the territory of Falun Great Copper Mountain, a UNESCO World Heritage Site

(WHS) since 2001, which offers more detailed insights into the complexities involved in these recent efforts in revitalizing and repurposing industrial sites while preserving their historical and cultural significance.

In Sweden and other Nordic countries, outdoor recreation has long traditions that promote the participation in outdoor activites by a large proportion of the population (Sandell and Sörlin, 2008; Hedenborg et al., 2022). Since the beginning of the 2000s, outdoor recreation has also become one of the political goals of the Swedish national government (Hedenborg et al., 2022). The growing demand, coupled with outspoken public policy for activity-based outdoor experiences, has extended to former industrial heritage sites, leading to their re-imagining as venues for new experienced-based products. Investments have been made in developing mountain biking (MTB), gravel biking and hiking trails, among other activities, within these sites. With the introduction of new forms of activities, the way visitors interact with these heritage sites undergoes a transformation. This not only impacts the physical structure of the site but also influences the interpretation of its tangible and intangible assets.

The shift towards activity-based experiences raises important questions regarding preservation, authenticity and ethics. In some mining districts, the change in demand for leisure activities presents an opportunity to create new products and to revitalize the industrial heritage (Berkenbosch et al., 2022). This chapter uses a specific case study of a mining site in the town of Falun (Sweden) to illustrate how the transformation of a mining heritage site poses challenges to the meanings associated with this place. It delves into discussion surrounding legitimization of postindustrial development, preservation and authenticity in the context of shifting visitor demands and the development of new experience-based products within industrial heritage sites. Our results indicate the industrial history being narrated in a simplified and almost sanitized manner, leaving no place for contestation or negativity. This study also shows that a mining site in Falun still needs to become a place where open discussions and dialogues with the local society are encouraged.

Making sense of mining heritage

In this part of our chapter, the literature on heritage and its role in place-making is explored, considering the changing demands of the evolving experience economy. Scholars, such as Laurajane Smith (2006), Dallen Timothy (2011) and Catherine Wanner (2016) among others have contributed to our understanding of heritage and its significance. Simultaneously, there is growing recognition of the diversity of perspectives on heritage and the ambivalence associated with acknowledging and respecting the diverse societies and changing conditions of traditional welfare state models, not least Nordic welfare states (Souillac, 2011; Kraenzle and Mayr, 2016). Indeed, this awareness emphasizes the importance of recognizing and embracing different views when interpreting the heritage of mining activities. Furthermore, these different perspectives can contribute to the creation of a more inclusive and democratic representation of mining heritage, in addition fostering a deeper appreciation for the meanings that are attached to these mining landscapes.

Furthermore, heritage can serve as a source of pride, allowing individuals and communities to celebrate their shared history and cultural identity (e.g. Donaldson, 2018; Bockelmann, 2023). It can also act as a platform for political expression advocating for specific causes or asserting certain rights (or in the example of Saami, representative's battle for their indigenous land against the planned expansion of hydro power stations in Norway during the 1970s or the recent protest in the northern part of Sweden against open-pit mining in Kallak). At the same time, heritage has the power to foster human connection by creating a sense of belonging and shared experiences. It can serve as a common ground for different individuals and communities to come together, exchange stories and memories, and thus forge connections among members of society (Donaldson, 2018; Bockelmann, 2023).

Even in cases where heritage is ambivalent or carries complex historical narratives and contested meanings of mining, it can provide a unique perspective that helps individuals and communities connect to a place (Jonsen-Verbeke, 1999; Bockelman, 2023). Supported by the examples given by Emily Donaldson (2018), the ambivalent heritage as a concept acknowledges the layers of history and the contested meanings, allowing for a more nuanced and critical understanding of the past. It invites all of us to engage with heritage in a reflective manner, encouraging dialogue and debate in finding our connection to a particular place on the map of mining heritage. In order for a heritage site to represent an opportunity for an open and inclusive dialogue for a wider community, a set of attributes needs to be created. What is necessary is to create not only a perception of an existing bond between the surrounding society and a heritage site, but through a set of embodied experiences to provide a sense of belonging. How can the challenges of the present-day society aiming at a constant distinguishing and positioning of itself be met by heritage sites?

In this chapter, we illustrate challenges connected to the representations of the industrial mining past and the possibility of modern uses of the mining landscape for the needs of the current society. We utilize our own in-depth expert knowledge of the processes leading to the creation of an industrial heritage site confined within the UNESCO's World Heritage Status in Falun, based on the history of more than 400 years of mining. We have both been part of the in-depth research and analysis of the transformation of the industrial mining site into a place for touristic experiences and a place of learning of the glorious past. We followed this process as early as 2005, just four years after the site gained its UNESCO World Heritage Status. Daniel was at the time doing research on the digital transformation of the interpretation of mining activities in Falun and other industrial sites in Sweden. Albina's research focused on the analysis of a process of institutional path-dependency and possibilities to innovate and integrate local communities into a process of new product development and alternative ways to the interpretation of mining heritage. The inquiries made by both authors through the analysis of perceptions of interviews with principal actors and the study of marketing of the efforts to promote this site as one of the elements of the contemporary landscape, both in order to create the awareness among its own citizens but also to attract external investments, helped us to reveal a specific postindustrial strategy for the mining heritage in Falun.

Moreover, Albina has been acting as Dalarna University's representative in the World Heritage Council (WHC), created on the base of WHS in Falun. WHC is led by the governor of the Dalarna province and consists of the legal bodies economically and legally responsible for the survival and preservation of the WHS in Falun. (Principal members of the WHC are Dalarna County Administration Board, Falun Municipality and the Foundation of Great Copper Mountain, which is a private entity owning the physical mining site and thus responsible for its everyday management.) Thus, we base our analysis on the deep situated knowledge of the setting connected to the representations of mining heritage, supported by the extensive empirical data set (more than forty qualitative interviews) created in the periods 2008–11, 2017–21 and 2020–2, and consisting of viewpoints expressed by different actors. As critical citizens of the Nordic welfare state system we also realize that not all intentions of these various interest groups have been fully realized and implemented. Recent developments aimed at merging the interpretation of natural and cultural heritage of this landscape, which has been a subject of continuous mining activities during the past 400 years, deserves our special attention and critical consideration.

Understanding the historical and practical (?) value of mining heritage in Sweden

Interest in industrial heritage in Sweden dates back to the turn of the twentieth century (Heldt Cassel and Pashkevich, 2011; Pashkevich, 2017). It was during this time that industrial companies, driven by the engagement of the Swedish industry elites, began developing museums, exhibitions and collections dedicated to preserving artefacts and buildings associated with the history of industrial production. These initiatives were often closely linked to specific companies and their representatives.

According to Nisser (1996), the development of industrial heritage in Sweden can be organized into different stages. In the early 1900s, the focus was on gathering industrial artefacts and displaying them. During this period, efforts were made to collect and preserve items related specifically to industrial processes and technology. This was replaced by a period of forming museums, which began with the establishment of the Stora Kopparbergs Bergslags Aktiebolags museum in 1922, also the same company which owned a Falun copper mine. This museum served as a model for subsequent industrial museums that were established elsewhere in Sweden. The primary focus was still on industrial processes and technology, with each museum centring its story around a specific company. These early initiatives laid the foundation for the development of industrial heritage in Sweden, highlighting the importance of preserving and showcasing the country's industrial history.

During the 1950s and 1960s, industrial archaeology emerged as a recognized field, marking a new era of industrial heritage development in the country. During this period, the focus shifted towards the preservation and protection of buildings that held historical value for the understanding of the development of industrial production in the country (Samuelsson, 2002). Industrial archaeology aimed to uncover and study the physical remains of industrial sites, such as factories, mines and mills, to gain insight into their historical significance. In 1973, a significant international conference was held at Iron Bridge, England, focused on industrial heritage, which sparked a broader movement across Europe. This event played a crucial role in initiating similar projects and initiatives centred around industrial heritage preservation in various countries.

The economic crisis of the 1970s marked the decline of the once-booming Swedish heavy industry (including mining and further refining of minerals), and structural reforms were implemented, resulting in deep changes to the national economy. The restructuring of global steel and raw material markets, coupled with the sharp rise in oil prices, had a particularly negative impact on heavy industry in central Sweden. The increased cost of energy eventually resulted in widespread unemployment in the Bergslagen area, and numerous industrial production facilities were forced to close. The consequences of these shifts in the world economy were substantial for the national economy, with an estimated four out of ten jobs in the Swedish metal and forestry industries disappearing after the 1970s (Lundmark et al., 2016).

In the 1970s and at the beginning of the 1980s, industrial heritage development witnessed a widening of interests and perspectives. It went beyond preservation of solely industrial environments and began incorporating other aspects of heritage, including worker's housing and workplace history. A large number of study groups were formed in many places around the country. The aim was to document and show the local industrial history (Isacson, 2013). This growing recognition of the need to tell a more comprehensive and inclusive story, encompassed not only technological progress but also social and cultural dimensions of industrialization represented by its workers (Nisser, 1996). Overall, the evolution of industrial heritage development in Sweden showcases the progressive expansion of its scope and objectives.

The development of an industrial preservation network in 1986 in the area of Bergslagen, which later transformed into the Ekomuseum Bergslagen, signifies the recognition and value placed on preserving the industrial heritage connected to the industrial archaeology mentioned previously. The establishment of an open-air museum instead of traditional indoor museums aimed to provide visitors with an immersive experience, by presenting historical artefacts in their original location. The development of the Ekomuseum Bergslagen and its focus on presenting industrial heritage in situ also helped to demonstrate a commitment to preserving and promoting the unique industrial history of this landscape. The selection of industrial heritage sites was usually made based on age, uniqueness and aesthetic criteria. The abandoned industrial estates from the 1970s were often taken over by municipalities and turned into light industrial or office estates, since the booming economy of the 1980s had created a demand for these buildings. They were not included in the industrial heritage initiatives to any large extent.

Indeed, the primary focus of Ekomuseum was to present a comprehensive view of the industrial processes connected to mining in the region. Education played a central role in this work. Although the specific focus and scope of the network has evolved over the years, it has remained in existence. What began as a collaboration between two municipalities in the southeastern part of Dalarna County gradually expanded to include more municipalities in the Bergslagen region. As of 2022, Ekomuseum Bergslagen comprises sixty-eight visitor sites dispersed throughout the Bergslagen area. This development can be seen as a means of strengthening the identity of the region, particularly during a period of economic restructuring and change. The partnership within this initiative facilitated the sharing of resources, knowledge and expertise, which promoted and sustained the industrial heritage of the region and also offered a range of experiences to engage with a wider audience.

Recent changes in perceptions related to industrial heritage preservation

During the 1980s, the Swedish economy underwent a significant transformation with a booming economy based on the liberalization and greater emphasis on commodification of products and services. This shift had implications for the industrial heritage sites in Sweden. Many industrial milieus became abandoned in the 1970s, resulting in vast, neglected industrial landscapes that conveyed a sense of hopelessness. Industrial preservation actions were, despite this, often directed towards older industrial buildings and not modern industrial complexes. The study groups that were organized in the 1970s lost momentum and the focus of industrial heritage work changed.

The transition period of the 1980s and 1990s, following mines and industrial sites closures in the 1970s and partial relocation of the industrial activities, resulted in a greater focus to nuance public understanding of mining heritage as a resource for the local economies. This time through service creation and leisure activities, and increasingly through the ideas related to product development for tourism, both domestic and international, especially following Sweden's joining of the European Union. This coincided with yet another deep economic recession at the beginning of the 1990s, which led to publicly funded labour market projects. These projects provided municipalities with the resources to restore and convert some of the previously abandoned industrial sites and to finance education in industrial history to raise awareness of the industrial past. There was a sense that this economic recession marked the end of the industrial era in Sweden, and that it was important to save the historic monuments from the industrial era. The combination of EU funding and publicly funded labour market projects created avenues for the revitalization of industrial heritage sites.

Since the early 2000s, commodification of heritage, driven by the need to make these sites attractive to tourists in order to justify public investments, has influenced the development of industrial heritage sites in the area of Bergslagen. Many industrial sites underwent renovations funded by the national state and the EU during this period (Pashkevich, 2017). One notable example is the establishment of a cooperative association called 'Intresseföreningen Bergslagen' in 2007, comprising public actors' representatives from municipalities, county administrative boards and regions. By 2016, it had twenty-three member municipalities, one county administrative board and three regions. The objective of this association was to use industrial heritage in the region as an infrastructure for fostering additional economic growth through entrepreneurship, innovation and tourism (Lundmark et al., 2016).

Several of these sites were included in cultural heritage routes, with the overarching idea to create various touristic experiences along these routes following the mining landscapes connected to exploitation of predominant minerals – iron, copper and silver (Iron Road, Husbyringen, Kopparleden and others) (Pashkevich, 2017). Looking back at this development today, it also became clear that the leaders in this process became sites that historically played the largest importance for mining or nodes of further refinement of iron or copper ore, thus having strong supporting economic structures, and knowledge and innovation fusion (Falun Great Copper Mountain, Sala Silver Mine in Sala, Tuna Hästberg Adventure mine outside the town of Borlänge, Dalhalla outside Rättvik, Avesta Art in Avesta, the Iron Works of Engelsberg). These were also the sites endowed with preserved industrial structures or buildings somehow related to mining or mineral production, and in addition, significant underground areas allowing for the development of immersive touristic experiences (physical visits to the underground mines).

This approach reflects the public policy stance at the time, where industrial heritage was viewed as a potential driver of economic development (Hospers, 2002; Heldt Cassel and Pashkevich, 2011; 2014). The public sector played a significant role in funding restoration and promotion of these sites, creating opportunities for further commodification of heritage sites and recognizing their potential to contribute to local and regional economic growth. Once again, mining was able to contribute to the local socio-economic development, but now through the development of small-scale entrepreneurship in leisure and tourism activities. However, evidence revealed that not all abandoned mining landscapes could be turned into successful tourist attractions, even if considerable investments were given to support these local initiatives (Pashkevich, 2017).

Creating the servicescape based on industrial heritage

In the pursuit of attracting tourists, there has been a shift in focus from educational activities to experiencing production within industrial heritage sites. Rather than solely preserving and restoring the built environments and collecting artefacts, the emphasis has shifted towards preservation and interpretation of heritage. According to the classical work of Pine and Gilmour (1998), the contemporary economy is driven by the production of experiences. These experiences encompass four dimensions: educational, escapism, entertainment and aesthetic. They can involve either passive or active participation, and can be immersive or absorptive for the consumer (Hayes and MacLeod, 2007). To be successful, managers of heritage experiences must develop products that offer visitors a compelling and engaging experience. In a study conducted in England by Hayes and MacLeod (2007), it was found that storytelling plays a crucial role in the success of heritage trails. The stories presented need to be written in a way that creates a coherent narrative for visitors, taking into account the diverse audience. Moreover, the study also revealed that most of the trails examined by researchers were designed for the general public, and there was a lack of visitor segmentation.

Conti's (2023) perspective on the experience economy brings attention to the interactive nature of experience development and the complex process of experience valuation. Although the traditional focus has been on the producer, Conti argues that the value of experience is created within the visitor through a multifaceted identity-building process. In recent years, there has been a shift in research towards understanding the consumer experience and the value that is derived from the interactions between visitors and the destination, both on-site and off-site (Conti and Lexhagen, 2020). This shift recognizes the importance of the visitor's role in co-creating the experience, and acknowledges that the value of the experience is subjective and individual. Within the context of industrial heritage sites, the influence of the experience economy paradigm has led to more activity-based offerings, where visitors actively engage and participate in the heritage experience. Visitors are encouraged to shape and personalize their experiences, contributing to the co-creation of the overall experience.

The shift towards visitor engagement and co-production of heritage experiences and the widespread use of smart technologies has indeed raised important questions regarding the sense-making of heritage. In a society where the linkage to the industrial production process is getting weaker, it becomes crucial to consider the lifeworld and background of visitors in the experience production process. This means that heritage experiences are expected to reflect the variety of understandings of the world that individuals carry with them throughout their lives.

Sweden, being a multicultural society today, presents a diverse range of visitor backgrounds compared to the initial stages of industrial heritage preservation in the 1980s. The diversity brings forth a variety of perspectives, cultural values and expectations when it comes to engaging with heritage. By taking into account the lifeworld of visitors, heritage experiences can be designed to resonate with their lived experiences, backgrounds and personal interests. This approach can potentially foster a sense of ownership and relevance, enabling visitors to connect more deeply with the heritage being presented.

Indeed, the co-production of heritage experiences brings forward questions regarding authenticity and ethics. Authenticity refers to the faithfulness and integrity of the heritage experience in representing the historical reality. In the early stages of industrial archaeology and heritage development, as we have discussed previously, the focus was on understanding and explaining the historical facts related to the industrial production process of specific industries. This approach aimed to provide visitors with an accurate and objective representation of the past. However, in the later stages of industrial heritage development in Sweden, there has been a shift towards a more interactive and participatory approach (Heldt Cassel and Pashkevich, 2014; Della Lucia and Pashkevich, 2023). The outcome of a heritage experience is no longer staged and predetermined by the curator, but becomes a result of negotiation between the visitor and the provider of this specific message. Thus, balancing the desire for visitor engagement and participation with the preservation of historical accuracy and integrity becomes a crucial challenge for industrial heritage sites. On the other hand, industrial heritage sites may actively challenge this notion and choose not to engage with alternative interpretations of their past because of external factors influencing them (e.g. UNESCO's World Heritage convention or inherited institutional settings) (Heldt Cassel and Pashkevich, 2011; 2014).

In the following section, we apply the suggested framework to the development of our case study of WHS Great Copper Mountain in Falun, thereby exemplifying the emphasis on the production of mining heritage experiences as they became known today.

Mining heritage of Falun

Falun is situated in Dalarna County in central Sweden (Figure 7.1). It is located approximately 300 kilometres northwest of Stockholm and lies in the northern part of Bergslagen, which is known as the traditional iron, copper and silver

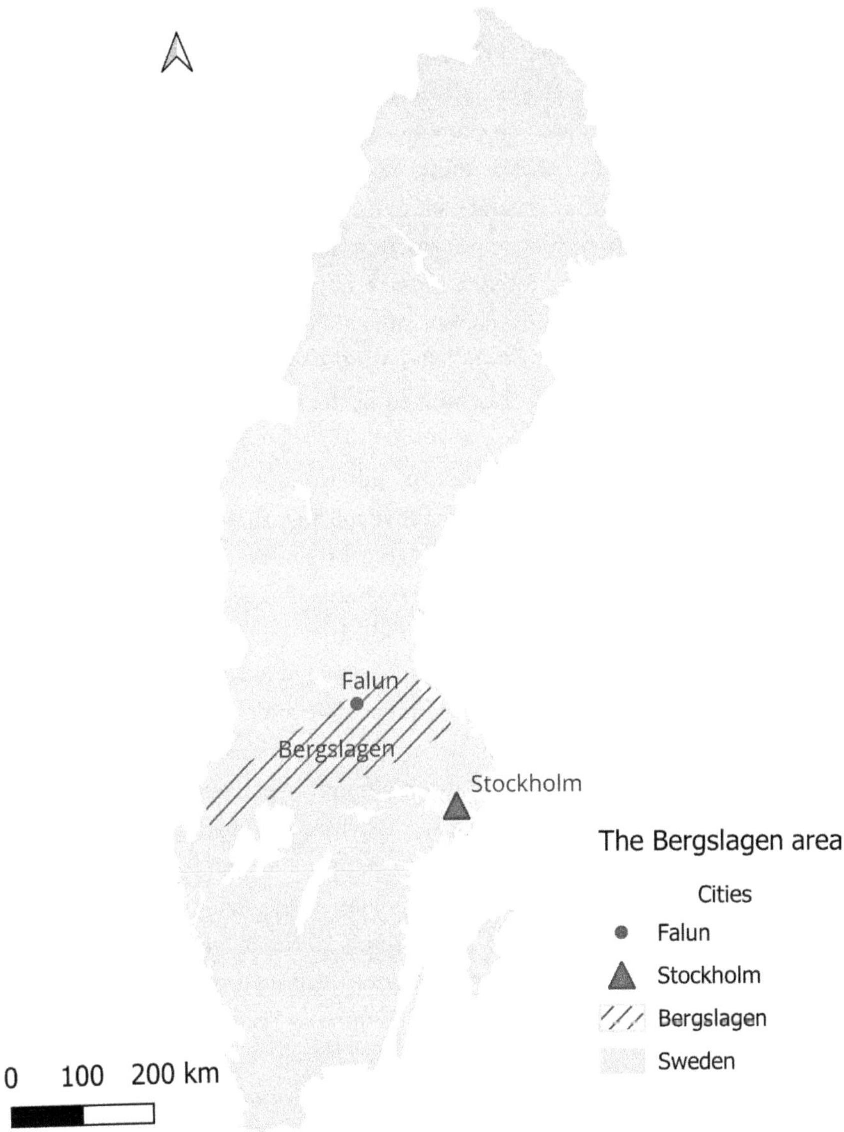

Figure 7.1 Map over the study area. Cartographers: Authors.

production and mining area in Sweden. The Bergslagen region has a rich history of metal production, and it is home to the oldest blast furnace in Sweden. The excavation of this furnace took place in the 1970s, and analysis indicates that metal production in the area dates back to the twelfth century (Magnusson, 1982).

The history of the Falun copper mine can be traced back to at least the eleventh century (Eriksson and Qvarfort, 1996). Over the following centuries, it grew in significance because of technological advancements, and it became a vital component of the Swedish economy. During the seventeenth century, it is estimated that the Falun copper mine accounted for two-thirds of the global copper supply (Riksantikvarieämbetet, n.d.). The mine played a crucial role in strengthening the Swedish state bureaucracy and financing wars, which led to the territorial expansion of Sweden in the sixteenth and seventeenth centuries. As the centuries progressed, the copper mine underwent transformations during the rapid industrialization of the nineteenth and twentieth centuries. The production at the mine was discontinued in the early 1990s, marking the end of its active operation.

According to the previous section of this work, it became clear that neo-liberal ideology had played a transformative role in reshaping the significance of Bergslagen's industrial heritage by realizing its commodity value. During the 1970s and 1980s, the focus on industrial heritage work was centred around restoration, and the preservation of industrial environments. Educational activities were emphasized, and visitors were guided through both above-ground and underground spaces (where possible and safe for the visitors). In Falun, a visitor mine was established in 1970, operating in conjunction with contemporary mining operations at the site until the closure of the modern mine in 1992. Over the years, tourism activities gradually evolved, but they received a much-needed boost after the inscription of the Falun copper mine onto the UNESCO World Heritage List in 2001, recognizing its well-preserved mine, workers' housing areas, and production facilities in and around the city of Falun. It also became known by its new name, Falun Great Copper Mountain, but by the public it was simply called 'Falun mine' or 'Falugruva' in Swedish.

Development of tourism and outdoor recreation spaces in the mining landscape

The mining landscape comprising the WHS in Falun serves as a classic example of an open-air museum (Figure 7.2). Furthermore, it also offers the opportunity for more immersive experiences through visits to the historical underground mining facilities. This is made possible by visits to the underground mine space located 60 metres below ground, with the help of the guided tours open to visitors all year round. (Tours are given in both Swedish and English, and the

Figure 7.2 The area of the Great Pit of the Falun Copper Mine (Photo: authors).

frequency of these tours varies according to the season). In general, the WHS as a tourist destination developed a seasonal character, as domestic and international tourists' arrivals are mostly confined to the summer months. In order to accommodate the influx of visitors, temporary staff, including guides and service personnel, are hired on a seasonal basis.

In addition, efforts are made to target international students from the local university who have proficiency in several foreign languages. Among international visitors to this part of Dalarna, the largest tourist groups traditionally come from Norway, Denmark and Germany, with a growing interest observed from the Netherlands. The popularity of winter tourism, particularly ice-skating on Lake Runn, played a significant role in attracting visitors from the Netherlands.

The network of actors connected to the cultural landscape of copper mining in Falun became one of the driving forces behind a successful nomination to UNESCO's World Heritage List (WHL) of the area of Falun Great Copper Mountain in 2001. It was not only the mining site that became part of the nomination, but also central parts of Falun town with its typical wooden villas and miners' estates outside of Falun in the local village called Staberg. The development following nomination can be summarized into two distinctive periods connected to the reintroduction and limited reinterpretation of the site, which lasted between 2000 and 2009, and the next period of reinvention and further reinterpretation of the mining landscape, which began in 2010 onwards.

The first period can be characterized by tourist experiences limited to only a few activities corresponding to the general trajectory of the development in line with UNESCO's guidelines (Della Lucia and Pashkevich, 2023). It mostly focused on physical visits to the historical underground mine and traditional mining museum. This attracted school groups and touristic visitation was confined predominantly to the period of June to August. Parallel to this development another local initiative started to evolve focusing on lived experiences through theatrical performances. These were aimed at exemplifying the life of ordinary citizens not involved in mining operations and took place during the summer months of the year. They were only available in Swedish. This innovative approach was taken by the provincial Dalarnas Museum in collaboration with the World Heritage coordinator and was financed by the Culture Unit of Falun Municipality (see a detailed analysis of this initiative in Heldt-Cassel and Pashkevich, 2014).

UNESCO's nomination of Falun Great Copper Mountain on to the WHL had already predefined a focus on the time period of the mine's heyday and its subsequent importance during the seventeenth century. Thus, interpretation focused specifically on this historical period of development, meaning other time frames of uses of the mine and its mineral deposits were omitted. However, this type of predefined and one-sided focus on the old times had consequences for the type of potential uses and possibility of further product development (Heldt-Cassel and Pashkevich, 2011; 2014).

The focus on the spectacular, unique and importance of mining in Falun for the industrial development of the whole nation, but also Europe's nation building, became one of the central underlying motives of storytelling and images in the marketing of the site to potential visitors (Pashkevich, 2017). Emphasis on the glorious past, a legitimate story of Sweden's greatness as a nation, continued to dominate visitors' experiences at the mine during this first period after the nomination. Furthermore, it also became evident that not enough attention was given to the role played by mining workers and the local society, or to the connection between mining operations in other regions and places within Sweden and outside, which contributed towards the international and national importance of this natural resource. Thus, narrating the industrial history in this sanitized manner omitted the role of people and left no space for contestation or conflict. In addition, an actual spatial separation among the three distinctive parts of the WHS has influenced the possibilities of their interpretation and considerably influenced their visitation. (For example, former mining houses even up until today are privately owned, thus a common strategy supported by the necessary infrastructure (e.g. guiding) was necessary in order to make them publicly available for visitation.) The focus of marketing material on only a few iconic facts has also contributed to the rather shallow representation of other aspects of mining development at the site during the 1600s to1700s. In general, a romantic and unproblematic picture of Sweden as a nation mobilizing its own natural resources in order to build a united Europe was put forth.

The mine is considered as a resource for commemorating the history of the past importance of the nation and its contribution to the build-up of Europe's economic power. It is a heritage that previous generations promised to preserve, and its importance needs to be remembered, meaning that preservation and conservation efforts will prevail. During this initial period, tourism development at the site in Falun has not significantly contributed to the preservation and further development of products and experiences, but instead there has been a reliance on the already established national and EU funds in order to protect these historical values. This period's main contribution is the establishment of a system aimed at preserving the site aligned to both national and European systems for heritage use. This period has also been marked by a growing sense of ambivalence towards this development by the local population, realizing how their modern needs are easily coming into conflict with the preserved heritage values.

The period after the WHS nomination and since 2010 was signified by changes in the managerial leadership of the Falun site. It also became apparent that the previous experiential practices had to be widened and adjusted allowing for different needs and adapting to the current society. This development led to a more nuanced approach to storytelling in guiding activities at the site. At the same time, the power dynamics related to the dominant character of the mine as one of the leading tourist attractions in the region made it difficult for local private initiatives to get involved in a possible collaboration with the site (Nisser, 2023). The perceived unspoken 'quality reassurance' procedure made it difficult to find ways to collaborate with the site, especially if your profile slightly differed from the perceived standard. This has resulted in several parallel developments on the territory of the site and in the city. According to local actors in Falun, if these efforts had been more coordinated from the beginning there would have potentially been a larger impact/contribution to the current entrepreneurial spirit of the city and consequently to the socio-economic development of the whole Dalarna County (Nisser, 2023).

Today, the mine site invites visitors to purchase high-end products at the mine's own souvenir shop, with the promise of taking back home 'a unique piece of Dalarna' or a 'Falun mine'. The shop specializes in selling examples of traditional, local food, and the site's own merchandise. Sustainability is highlighted as a necessary selling point, and it is mentioned that all products are created using sustainable and environmentally friendly practices. Other local shops in the same area are also connected to the theme of products made by hand, either pieces of art or fine food, adding towards their exclusive and high-end character.

Development of outdoor activities

A significant part of the development after the nomination of Falun mine to UNESCO's WHL was the development of possibilities to increase secondary spending at the site and trying to make sure that visitors remain at the site as long as possible. Consequently, the outdoor environment also needed to be further adjusted to create the possibility for walking or biking around the Open Pit area (the part of the mine that collapsed during the mid-summer celebration of 1687). Nature also received greater attention in 2010, when a parliament of eagle owls settled at the mine. A project financed by the Swedish Environmental Protection Agency was organized by the mine management to protect the eagle owls and signs were introduced containing information about eagle owls. Telescopes were also placed in observation towers in the open pit area. This project shows the ambition of Falun site's management to help to further develop the mining environment and to potentially attract visitors interested in birdwatching to the site.

Certain adjustments made in the outdoor environment of the landscape directly connected to the mine resulted in the creation of an almost 30-kilometre-long hiking trail, which signified the importance of access to the waterways previously used for mining activities. The physical map is made available for pickup at the Falun mine visitor centre. The map shows marked entry points, parking possibilities for cars and camper vans, places to rest along the way and in addition some information boards explaining what visitors can see in the landscape.

Cycling has grown in popularity in Dalarna and traditionally this part of Sweden is well-known for its MTB trails and active biking clubs. Local Destination Management Organisation and tourism businesses have been a part of several EU-financed projects to develop this kind of activity. Information about this path is available for visitors on the naturkartan.se site, where Sweden's municipalities are able to publish information about hiking, biking and MTB trails, and other nature-based activities present on the site (Naturkartan, n.a.). It is interesting to note that among the few comments from Swedish-speaking visitors available on this site who have experienced this trail, some have stated that it is lacking important markings and thus visitors have difficulty understanding how long they have left to go on the marked route they have chosen. These were native-speaker hikers as well, which would mean that the focus has not been to interpret the cultural heritage of mining for the wider group of visitors, including foreigners.

Another interesting observation was also made based on comments from users that the cultural heritage of the mining industry values was secondary to the nature-based experiences. Therefore meaning that visitors choose this area not for its cultural values, but for physical activities, such as hiking, as it is located in close proximity to an urban area and is thus accessible. Two years have passed since the publishing of the first reviews on the site devoted to these nature-based experiences and the last comment was published in the past eight months. There is no feedback provided to users, thus making this less interactive, and it is unclear how impressions and suggestions from the first-hand users are reaching the Falun Municipality, which is formally responsible for the upkeep of these nature trails.

Yet another possibility is a hiking trail developed specifically for families with children in Staberg village. These are examples of miners' living quarters with baroque gardens, including a restaurant and a coffee shop. These activities are 1.5- and 2-kilometre long trails, where children can follow two historical characters on an exciting adventure in the landscape that is a part of the WHS. At the beginning of each path you can get a map and follow a thematic stroll in the landscape.

The transformation of the Falun mine area into an arena for other outdoor activities apart from hiking has been facilitated in two ways. First the WHS is made available to start-ups specializing in outdoor experiences, such as companies organizing standup paddleboarding (SUP) in the waterways connected to the mining and mountain bike tours and similar activities. Secondly, the area is also used for sport events, tapping into the expanding public interest for physically demanding and adventurous activities in combination with organized sport events. Falun is a well-established sport and event city. The local multisport arena, Lugnet (a famous national arena for cross-country skiing in Sweden) is located at the other side of the town, and hosts numerous sports and other events. Because of this, it attracts many visitors that stay at the local hotels and campsites. The connection between Lugnet and the World Heritage Site has been weak according to opinion's expressed by several actors included in a recent study of the possibilities and potential for sustainable tourism development of the industrial heritage of the site (Nisser, 2023). To attract more visitors to the heritage site, new sports events have been developed in the heart of the heritage area. One example is the running event that has been hosted annually since 2018, emphasizing the area of the Great Pit at the heart of the mining site in

Falun (Storastotenrun, 2023). It is a competition that leads the runners into the former mining area and follows the waterways that were originally created for powering the pumps and hoists in the mine before it was electrified.

The running event is also using a previously created walking trail as an arena for this new outdoor experience. On the event website, the mine is marketed as a milieu that has been formed by the hard work of people connected to the mine. Runners will pass by old iron works and dams and canals leading to the mine. The website gives a brief mention of the historical event connected to the collapse of the roof of the copper mine in the 1600s, which formed an area currently known as the Great Pit. It highlights the fact that no one was killed in the accident because it was a public holiday (Storastotenrun, 2023). The mining heritage is used as a milieu where other popular activities can be performed without having to re-develop new ways of interpreting the local history. This is an example of how the industrial heritage site is being re-branded as an outdoor arena that draws on the history of the place.

Reflections and ways forward

The case of the Great Copper Mountain in Falun, Sweden, shows how industrial heritage is narrated for the sake of the demands provided by a consumer-driven developmental strategy of tourism development. Efforts to open up the possibilities for outdoor experiences in the cultural landscape of mining are following a general trend that became especially evident since the beginning of the 2000s, further highlighted during the COVID-19 pandemic with the increased interest in domestic nature-based activities while international travel was still on hold.

Nowadays, even the natural environment surrounding the mine is slowly becoming commodified. The connotations of humans' own power to form their bodies and mind while exercising and taking part in sports events are used in order to be compared to the former uses of nature by humans to extract natural resource wealth from the copper deposits dating back as early as the twelfth century. It becomes self-evident that humans are in control of the processes forming not only their own bodies while running, hiking or biking, but also gives them full right to decide on how to form a natural landscape. This new use of the mining landscape as an arena for outdoor experiences is formed by the consumption needs of the Swedish society today. Thus, a site becomes yet another setting allowing this postmodern consumption to situate itself in the modern landscape (Berkenbosch et al., 2023; Della Lucia and Pashkevich, 2023).

The self-guided hiking trails and biking are also developed to correspond to the needs of the present-day consumer, and the industrial heritage acts as a background. The narration allowing visitors to understand and appreciate the history of the site is still largely analogue and has not been transferred or transformed to suit these new uses of the area to become more connected and virtual. Thus, the link between nature-based and heritage experiences is missing, and visitors are expected to actively express their interest in understanding a history of the place.

Once again, this given example illustrates how the commodification of history of the industrial use of the area proceeds to suit the needs of the postindustrial region, helping it to strengthen its position in the economic competition for labour and investments. This process is helping to create an illusion of an extended use and showcase 'new' interpretations of the landscape by trying to bridge parts of its industrial history with a focus on nature. However, it is only a matter of segmented and confined groups in society being able to use the space for outdoor activities. There is still a lack of support towards new types of narration. The possibility of expanding the customer group is still limited and the competition from other adventure companies offering similar products and experiences is also rising.

This overview was able to connect the recent history of the successful industrial heritage preservation and still ongoing interpretation process in Sweden. We exemplified this process by using an example of one of the largest mining provinces in the country – Bergslagen and the town of Falun with its Great Copper Mountain – in order to shed light on the present stage of uses of mining heritage. This brief analysis of the present-day outdoor leisure uses of the mining landscape in Falun gives us an opportunity to observe what is still missing and what potentially needs to be adjusted in the work towards ascribing new values to old industrial spaces.

The focus on the creation of spaces for outdoor activities has become a recent focus for the institutions involved in both preservation and utilization of the WHS in our case. The interpretation of the historical past is also shifting from being oriented on facts and information towards a lighter version supporting the demands from the postmodern consumption culture (Conti, 2023). This results in a segmented, more shallow representation and narration to visitors. At the same time, we need to be reminded that a shift towards the physical use of historical mining areas has direct implications onto its tangible values. The shallow representation of both cultural and natural values of mining heritage leads to a lesser appreciation and understanding of the impact created by mining activities for Sweden's role as an industrial economic society and later on as an example of postindustrial nations. Current narrations of mining heritage risk simplifying and omitting most of the negative aspects related to the mining operations in terms of their environmental and societal impacts.

It is highly relevant to look closely at how the creation of new products and experiences are perceived by the customer groups that are specifically targeted by these activities, but also to go beyond them and see how the current mining landscape is consumed and understood by Falun's own and Dalarna county inhabitants. Currently, an appreciation of the industrial history of the World Heritage Site and its values is possible only by active engagement with the heritage via educational programmes provided to the elementary schools in the county of Dalarna and administered by the site in Falun, and via the direct visits to the mining museum and the underground mine. Other types of outdoor activity consumption at this space are not supported or led by any expert interpretation, but instead rely on simplified and brief mentioning of the spectacular facts (for example with the case with the Great Pit run).

Another interesting remark and a possible line of future enquiries is the possibility of the establishment of a link between the short- and long-term impacts on the natural environment created by mining operations. This relationship is absent at present and focus hereon lies in the narration connected to cultural and economic impacts confined to mining heritage. The mining site in Falun has yet to explore its potential in order to influence its visitors' awareness of the enormous and long-lasting impacts of mining as an industrial activity. It is especially important when this region is already witnessing the direct consequences of climate changes – an increased frequency of severe spring flooding and increased precipitation throughout the year, which represent a serious threat to the tangible heritage not only directly preserved by the WHS site in Falun, but also the tangible heritage of parts of the city that are not necessarily given the same high status.

References

Berger, S. (2019), 'Narrativizations of an Industrial Past: Labour, the Environment and the Construction of Space in Comparative Perspective', in S. Berger (ed.), *Constructing Industrial Pasts: Heritage, Historical Culture and Identity in Regions*

Undergoing Structural Economic Transformation, 288–305, New York and Oxford: Berghahn Books.

Berkenbosch, K., P. Groote and A. Stoffelen (2022), 'Industrial heritage in tourism marketing: legitimizing post-industrial development strategies of the Ruhr Region, Germany', *Journal of Heritage Tourism*, 17 (3): 327–41.

Bockelmann, L. (2023), 'Impacts of Change: Analysing the Perception of Industrial Heritage in the Vogtland Region', *Urban Planning*, 8 (1). https://doi.org/10.17645/up.v8i1.6025.

Conti, E. (2023), 'Digital Technocultures in Nature-based Tourism', dissertation, Sundsvall: Mid Sweden University.

Conti, E. and M. Lexhagen (2020), 'Instagramming nature-based tourism experiences: A netnographic study of online photography and value creation', *Tourism Management Perspectives*, 34: 100650.

Della Lucia, M. and A. Pashkevich (2023), 'A sustainable afterlife for post-industrial sites: balancing conservation, regeneration and heritage tourism', *European Planning Studies*, 31 (3): 641–61.

Donaldson, E. C. (2018), 'Place, Destabilized: Ambivalent Heritage, Community and Colonialism in the Marquesas Islands', *Oceania*, 88 (1): 69–89. https://doi.org/10.1002/ocea.5182.

Ek, A. S., S. Löfgren, J. Bergholm and U. Qvarfort (2001), 'Environmental Effects of One Thousand Years of Copper Production at Falun, Central Sweden', *AMBIO: A Journal of the Human Environment*, 30 (2): 96–103.

Eriksson, J. A. and U. Qvarfort (1996), 'Age determination of the Falu copper mine by 14C-datings and palynology', *GFF*, 118 (1): 43–7.

Funk, M. and A. Pashkevich (2020), 'Representations of industrial heritage in tourism materials: Analysing androcentric discourse in textual and visual content', *Dos Algarves: A Multidisciplinary e-Journal*, 36: 41–58. doi: 10.18089/DAMeJ.2020.36.3.

Hayes, D. and N. MacLeod (2007), 'Packaging places: Designing heritage trails using an experience economy perspective to maximize visitor engagement', *Journal of Vacation Marketing*, 13 (1): 45–58.

Hedenborg, S., P. Fredman, A. S. Hansen and D. Wolf-Watz (2022), 'Outdoorification of sports and recreation: a leisure transformation under the COVID-19 pandemic in Sweden', *Annals of Leisure Research*, 27 (1): 36–54. https://doi.org/10.1080/11745398.2022.2101497.

Heldt Cassel, S. and A. Pashkevich (2011), 'Heritage Tourism and Inherited Institutional Structures: The Case of Falun Great Copper Mountain', *Scandinavian Journal of Hospitality and Tourism*, 11 (1): 54–75.

Heldt Cassel, S. and A. Pashkevich (2014), 'World Heritage and Tourism Innovation: Institutional Frameworks and Local Adaptation', *European Planning Studies*, 22 (8): 1625–40.

Hospers, G. J. (2002), 'Industrial Heritage Tourism and Regional Restructuring in the European Union', *European Planning Studies*, 10 (3): 397–404. https://doi.org/10.1080/09654310220121112.

Isacson, M. (2013), 'Industriarvets utmaningar: Samhällsförändringar och kulturmiljövård från 1960-tal till 2010-tal', *Bebyggelsehistorisk tidskrift*, 65: 17–36.

Jonsen-Verbeke, M. (1999), 'Industrial heritage: A nexus for sustainable tourism development', *Tourism Geographies*, 1 (1): 70–85. doi: 10.1080/14616689908721295.

Kraenzle, C. and M. Mayr (2016), 'From Europe's Early Iron Age to a New Urbanist *Shanzhai* Village: Themed Environments, Global Property Markets, and the Role of Hallstatt's Cultural Lineage', in C. Kraenzle and M. Mayr (eds), *The Changing Place of Europe in Global Memory Cultures*, 201–19, Palgrave Macmillan Memory Studies, Cham: Palgrave Macmillan. https://doi.org/10.1007/978-3-319-39152-6_11.

Lundmark, M., M. Hedfeldt and M. Jakobsson (2016), *Bergslagen: en industriregion i upplösning?*, Stockholm: Riksantikvarieämbetet.

Magnusson, G. (1982), 'Lapphyttan: en medeltida hyttanläggning', *Kulturmiljövård*, 7 (2): 27–39.

Naturkartan (n.d.), 'Vattnets väg till gruvan'. https://www.naturkartan.se/sv/dalarnas-lan/vattnets-vag?utm_source=naturkartan&utm_medium=map (accessed 27 August 2023).

Nisser, M. (1996), 'Industriminnen under hundra år', *Nordisk Museologi*, 1: 73–82.

Nisser, U. (2023), 'Stärka världsarv i Världsarvet Falun 2019–2021: Projektsammanfattning', Region Dalarna, Kultur och bildning. ISBN: 978-91-527-0275-8.

Pashkevich, A. (2017), 'Processes of Reinterpretation of Mining Heritage: The Case of Bergslagen, Sweden', *Almatourism, Journal of Tourism, Culture and Territorial Development*, 8 (7): 107–23.

Pine, B. J. and J. H. Gilmore (1998), 'Welcome to the Experience Economy', *Harvard Business Review*, 76 (4): 97–105.

Riksantikvarieämbetet (n.d.), https://www.raa.se/evenemang-och-upplevelser/upplev-kulturarvet/varldsarv-i-sverige/alla-varldsarv-i-sverige/falun-och-kopparbergslagen/ (accessed 9 August 2023).

Samuelsson, J. (2002), 'Industriarvets födelse', *Historisk discreet*, 122 (2): 2–6.

Sandell, Klas and Sörlin, Sverker (2008), 'Inledning', in Klas Sandell and Sverker Sörlin (eds), *Friluftshistoria: Från härdande friluftslif till ekoturism och miljöpedagogik*, 10–15, Stockholm: Carlssons Förlag.

Smith, L. (2006), *Uses of Heritage*, New York: Routledge.

Souillac, G. (2011), *The burden of democracy: The claims of cultures, public culture, and democratic memory*, Maryland: Lexington Books.

Storastotenrun (2023), 'Terränglöpning mitt i värdsarvet', https://www.storastotenrun.nu/ (accessed 27 August 2023).

Timothy, D. J. (2011), *Cultural Heritage and Tourism: An Introduction*, Vol. 4, Bristol, Buffalo and Toronto: Channel View Publications.

Wanner, C. (2016), 'The Return of Czernowitz: Urban Affect, Nostalgia, and the Politics of Place-making in a European Borderland City', *City & Society*, 28 (2): 198–221.

8

What the Muggles Do Not See: Affordances of Abandoned Industrial Sites for Hobbyists

Tiina Äikäs, Marjo Juola, Oula Seitsonen and Gabriel Moshenska

Introduction

Entering the outskirts of a derelict industrial area in the gloomy late afternoon light and light drizzle, one is acutely aware of the surroundings with the black, empty, broken-glassed windows staring down at you from the echoing red brick walls. We know that various activities still take place within the old halls and the area is not off limits, even if nobody is to be seen, and we are searching for an official geocache. Visiting the industrial area is part of our research into geocachers' perceptions on engaging with heritage and historical sites, but still the feel of intruding and of someone watching you is overpowering. It makes one instinctively hush their voice and footsteps to move with stealth and caution to avoid attracting attention, while following the hints given by the cache owner to locate the 'hidden treasure' hiding in plain sight, camouflaged so that only the initiates can find it and see the hidden potential, something that the muggles won't see.

Research diary entry by Oula Seitsonen from October 2018 of a visit to a semi-abandoned industrial area in Lieksa, eastern Finland

Deindustrialization is a powerful concept to think with: it speaks to globalized supply-chains, impoverished communities, and – most iconically – a titanic infrastructure of factories, refineries and pit-heads fallen silent. These remains are commonly described in terms of abandonment, emptiness and wastelands. But to define such sites solely in terms of their industrial past is to deny them their present and futures. Old industrial sites can seem to lie in wait for something – for new uses and re-uses, for 'heritagization' or for destruction. Even in an indeterminate, intermediate state, awaiting the next chapters of their lives, they can play powerful and significant roles for people's sense of place and their engagement with their material and spatial surroundings. This study sets out to explore how hobbyists and other users choose to engage with old industrial sites as lived, experiential places. The diary entry that we started with, of a visit to a semi-abandoned industrial site in eastern Finland, echoes in many ways the themes and sentiments expressed by the hobbyists that we have interviewed.

The abandoned Finnish industrial sites and spaces that we discuss below represent different levels of abandonment. They are mostly situated in the southern half of Finland in the towns with old industrial communities. Some of them have been officially completely abandoned, while at some of the sites a few activities still take place in parts of the industrial complexes, for example, artistic or handicraft workshops. In some cases, these include illegal or illicit activities, such as a motorcycle club and a 'drug factory' at one of the sites. None of the places that we deal with are recognized as heritage sites, and they are at different stages of ruination (Figure 8.1). All the sites still have standing structures, and many of them also have differing amounts of machinery and other industrial material culture present. The abandonment, neglect, (un)official re-use and leftovers of industrial material culture all contribute to the special character of these spaces.

Figure 8.1 One of the visited sites is an old brick factory in southern Finland (Photo: Oula Seitsonen, 2022).

For this study, we approached practitioners of several different hobbies that share an active bodily engagement with their environment:

- **Parkour** is a discipline based on athletically and artistically overcoming urban obstacles, traditionally as a way of finding the straightest way between two places: hence sometimes using parts of the built environment where people do not usually go.
- **Urban explorers** (UE) enter abandoned and empty buildings, aiming to 'leave nothing but footsteps and take nothing but photographs' (Ninjalicious, 2005: 20).
- **Geocaching** seeks to locate deliberately concealed and camouflaged caches in urban or natural environments with the help of GPS coordinates and the geocaching application.
- **Airsoft** is a battle game where players, usually organized into teams, carry out missions armed with realistic-looking airguns that shoot plastic balls.
- **Urban survivalism** aims to develop and teach skills to help people survive in the aftermath of a major disaster or breakdown of society.
- **Rescue dog training** takes place in different environments including in simulations of different built environments, where dogs are trained in urban search and rescue.

What all these activities have in common is the use of industrial spaces in distinctive and unintended ways, i.e. finding ways of use for which the space was not originally intended (cf. Kidder, 2012: 230). Ameel and Tani (2012a: 164) have commented how 'practitioners of parkour (traceurs) build unique and highly embodied relationships with their daily space'. One of our interviewees describes parkour as 'being in interaction with one's environment' (Interview 65). The other activities outlined above also give practitioners new embodied perspectives on their spatial environments (cf. Ameel and Tani, 2012b: 17; De Martini Ugolotti and Genova, 2023: 3). Signe Højbjerre Larsen (2021: 26) depicts this as 'material curiousness' or 'material engagement', where the practitioners take a curious and imaginative stance towards their surroundings. By exploring these interactions, we can gain new knowledge of the meanings of abandoned industrial spaces.

This chapter draws on the findings from eleven semi-structured interviews conducted between 2021 and 2023 as a part of the 'Smokestack Memories' project (see Preface of this book for more details), which were focused on recreational use of the spaces or touched the topic. Other interviews were aimed at different research questions, e.g. working at the sites or living close to them. The age range of the selected interviewees varied from nineteen to fifty-one, and

Table 8.1 Interviewees cited in the chapter

Interview number	Hobby	Age	Gender
59	UE	42	male
34	UE	38	male
60	UE	25	female
Bunnyurbex/66	UE	20–30	female
67	UE	*c.* 40	female
61	urban survival; UE; geocaching	29	male
62	airsoft	51	male
9	airsoft	19	male
65	parkour	43	male
63	parkour	35	male
64	rescue dogs	43	female
Anonymous 1	geocaching	41	female
Anonymous 2	geocaching	41	female

the gender distribution was rather equal (Table 8.1). Most of the interviews were conducted 'on location', providing an additional dimension of personal experiences of the sites and their atmospheres. Even though unauthorized visits are often part of some of these activities, for all of our work we acquired the necessary permissions and even had a property-owner as a guide. This of course affected the way we experienced the site, removing some of the danger.

We identified our interviewees via social media accounts and through contacts from earlier interviews. Our aim was to find people using old industrial sites as places for their recreational activities. It soon became obvious that this self-identification was not consistent across all the groups. Two of the people who we interviewed in relation to UE declared that they did not consider themselves as urban explorers: one stated that the whole term is stupid and that he and his friends preferred to use the Finnish verb 'koluta' (scour or ransack) (Interview 59). For the purpose of this study, the ways in which people use industrial sites are more important than whether they regard themselves as belonging to a certain group. We asked the interviewees how they used the space, why they selected specific locales and how they experienced the space. For this chapter, we analysed in more detail the responses that highlighted the affordances of industrial sites for these recreational activities. Our aim was to study how different hobbies enable people to see affordances in industrial spaces, and how industrial spaces enable different activities outside their original designed intent. This helps us to see how old industrial sites lie somewhere between wasteland and heritage, where their state of ruination can be a precondition for interaction.

Affordances for bodily engagement and imagination

> The hobbyist's own creativity and imagination are the main factors when it comes to places.
>
> Interview 63

'Affordance' is a concept that is used widely in a variety of fields from psychology to cultural geography and design. It refers to action-related properties of the environment (Heft, 2010: 28). We take as a starting point Gibson's ([1979] 1986: 127) definition that: 'The *affordances* of the environment are what it *offers* the animal, what it *provides* or *furnishes*, either good or ill.' Even though affordances are possibilities for action provided by the environment, they are not universally available. We follow Rietvelt's and Kiverstein's (2014: 325) argument that 'the affordances an environment offers to an animal are dependent on the skills the animal possesses'. Environmental psychologist Marketta Kyttä (2004) emphasizes that affordances consist of direct perception and actualization. An individual can perceive that the environment offers potential for activity, but this activity can only be actualized if the person in question has fitting physical abilities and personal intentions (Kyttä, 2004; see also Raymond et al., 2017: 7).

On this basis, industrial spaces that for some people seem to offer affordances only for bulldozers, may for skilled practitioners of parkour or geocaching offer venues for bodily engagement with the space. Edensor et al. (2011: 67) have described the affordances of ruins as a 'crucial quality of ruins is their material affordance, to be found within the unfamiliar, unkept environments that foster a multitude of opportunities for playful interaction with space and matter'.

One could argue that this kind of playful and imaginative interaction is an affordance that we humans as playful beings, *Homo Ludens* (Huizinga, 1938), are skilled in bringing to actualization. Historian Johan Huizinga has seen the instinct for play as the central element in human culture and it can guide our ways to interact with the environment. This widens the set of affordances from the motor skills and environmental resources that have often been the main focus of affordance studies, to include sociocultural practices and emotional opportunities typical for us as human beings (cf. Kyttä, 2004: 181; Rietvelt and Kiverstein, 2014: 343). As atypical spaces for activities of any kind, old industrial sites create affordances for new and imaginative bodily engagement with the environment and the space.

The thrill of exploration

The remote locations of many old industrial sites allow exploration to be carried out undisturbed, and explorers are often willing to take risks to gain access, as one of the interviewees stated: 'Well, there was quite an exciting power station, which was fairly well closed, so you needed to find a bit of a cunning route to get there through roofs, so it was of course a very exciting place, a place like that' (Interview 59). There is a feeling of freedom in entering spaces that one is not supposed to go to (Hudson, 2014: 196). One of our interviewees noted that in UE, he was mainly interested in old industrial areas that had not been converted or re-used. He wanted to explore unused and so-called 'abandoned' places. At the same time he was critical towards the concept of abandonment, stating that it was not a good word as 'you do not abandon buildings because they always belong to someone' (Interview 59). The attitude to abandonment was one of the central ambivalences for hobbyists as the abandoned nature of the sites enabled their practices, but many of them were at the same time aware of the possible owners of the sites.

This need to explore was shared by urban explorers and parkour practitioners. According to an interviewee, in its early stages, parkour came close to urban exploration as the practitioners explored places where people normally would not go or even have access to. Practising parkour gave traceurs the ability to go to these places using their skills and techniques to climb over walls. The possibility of actualizing these affordances had 'kept alive a childlike exploration [of spaces]' (Interview 65). Another traceur described exploration as a central aspect of parkour or even of the nature of the traceur, which was to explore and seek adventure and search for new places: to 'scout' (Interview 63).

Scouting could also include aspects of research as the practitioners of UE studied literature on industrial history and used Google Maps to record potential and already visited sites. The joy of finding and collecting were seen as similar to birdwatching. Three of the interviewees mentioned that getting background information on the sites had gained importance as they became more experienced (Interviews 34, 59, 60). One of them shared old photos and histories of the sites on Instagram. Background information could be collected either before or after visiting the site, and it was described as giving depth to the visit. 'You are able to better experience the presence of history when you know what has happened there and see to which state it has turned to' as noted by one interviewee (Interview 60).

Similar practices were also observed amongst geocachers, who often provide abundant background information of the cache sites in their descriptions. In fact, a tentative study that Seitsonen carried out with his colleagues some years ago in a Digital Humanities Hackathon showed that geocachers often place their caches at or in the immediate vicinity of heritage sites (DHH17, 2017; Immonen, 2022: 50). A caching background story can in some cases attract people to visit specific cache sites and explore places such as old industrial sites where they would not otherwise enter because of the sense that they have a threatening aura (Anonymous 1).

However, one interviewed urban explorer commented on the loss of a 'sense of exploration' as 'all information is already on the Internet' (Interview 59). In his view, urban exploration was not the same as before, as Facebook groups and the Internet in general share too much information. He compared it to the joy of finding the information in the 1990s and early 2000s when one got bits and pieces of knowledge from books and conversations. Compared to this, geocaching offers a kind of sanitized and safe thrill of exploration: geocaching has a set of rules, e.g. about placing the caches only in places where everyone can legally enter.

The remoteness, abandonment and secrecy of a place afforded a sense of exploration and research but also helped to protect the site, as one of the interviewees stated. The sites that were more remote were typically less vandalized and left 'in the condition in which it had been originally left' (Interview 34). Here the abandonment offered a sense of authenticity, the feeling of entering a place of the past that was untouched.

Exciting danger

> Visiting industrial sites evokes both excitement and special alertness as one knows that one is often there without permission.
>
> Bunnyurbex

Some of these abandoned places have also had more 'controlled' uses for urban survivalism courses and military training. An interviewee who had participated in an urban survivalism course at an old, abandoned lime and cement works called Kalkki-Petteri described it as a 'suitably vile place', 'ruthless' and 'impressive' (Interview 61). For this kind of course, where you are meant to learn to survive in extreme urban conditions, the −15 degrees Celsius and a space built of concrete was seen as a good combination. This is an example where normally negative aspects of space are turned into affordances for exciting and even purposefully dangerous, but primarily educational survival experiences.

At the survival course, the elements of danger were nevertheless controlled, and organizers are required to have valid insurance. As climbing accidents have been known to occur, the participants were also warned to look where they stepped, and parts of the building were closed. Some interviewees also emphasized that in practicing UE or parkour they did not purposefully seek danger. As one of them stated: 'I don't get my kicks off it, or specially care of the sense of danger, or even like it [the danger], but it just is what it is' (Interview 34).

But the sense of danger was nevertheless present in the abandoned and partially destroyed spaces, in part as a source of interest and affect. One interviewee compared Kalkki-Petteri in its shattered condition to the images of war-damaged buildings in Ukraine. Adrenaline was often mentioned in connection to visiting abandoned industrial sites. Some of the fears were described as irrational, such as 'you never know if there is an animal or a person who still lives there or you find a dead body or something' (Interview 60). These kinds of feelings have been described also by geocachers: even if their hobby offers a more curated and controlled version of exploration, entering dark factory halls, cellars and other (semi-)abandoned structures gives a rush of adrenaline and an acute awareness of one's immediate surroundings (both Anonymous 1 and 2). 'It feels like someone is peeping from behind your shoulder, just about to tap it', as one of them described (Anonymous 2).

Seeking adrenaline and excitement seem to be common but often unacknowledged aspects of these activities. One of the interviewees observed that adrenaline gives you a good feeling, and that others might want it, but claimed that it was not his purpose for visiting the places, even though one could feel relaxed and energetic after the experience (Interview 59). Another interviewee noted that these fears can sometimes be so overwhelming that it becomes impossible to open a door for fear of finding something hideous. But at the same time this feeling is a part of the appeal, and what drives the desire to return (Interview 60). Industrial sites in particular can evoke feelings of danger with their sheer size and monumentality. For instance, vast, echoing factory halls with a maze of ruined machinery easily spark a fear of the unknown and summon up impressions of similar sites in popular culture, such as crime and horror movies (Anonymous 2). The stage of abandonment with rust and unfamiliar metal structures seems to signal that it can be dangerous to go to certain places. Some interviewees noted that the aesthetics of these ruined industrial places is similar to those in horror movies and creates the impression that these places are no longer meant for humans, without risking fatal danger (Anonymous 2). One traceur describes this sense of danger in a very auditory way:

> What creates the atmosphere is that there is often piping and other things that make noises. That is exciting. You know that no one is moving there, no one is interested or they think it is dangerous but still you feel a bit tense when you hear bangs and the structure is living its own life. There might be some ventilation hatches or something that clanks [...] And then maybe only the dust that settles there if you go to a space where not many people have been. And about the noises one more thing, that often if it is a totally abandoned [building], there is glass or something else under your shoes that might make sounds. It creates a specific soundscape.
>
> Interview 65

The dark corners of abandoned industrial spaces were described as creating a sense of adventure as well (Interview 63). Light, or the lack of it, was also vital when urban survivors stayed for a couple of nights at an old lime and cement factory. The wide, open spaces of industrial estates were not always ideal for training. The participant in the urban survivalism course noted that Kalkki-Petteri was actually quite unsuitable for the activities because it gave no cover, and you could be observed from every direction. He describes how his imagination ran wild as he was standing alone in the middle of the factory hall looking through the dark doorway: outside the moon was shining but inside it was totally dark. He remembers thinking that if someone came now, he would lose (Interview 61).

Parkour is often associated with risk taking and danger and even turning safe environments into dangerous ones (Kidder, 2012: 238). One interviewee described risks and risk control as key elements of parkour. Similar to UE, the risk taking or excitement seeking might sometimes be denied, but our interviewee regarded this as hypocritical: risks might be small, but they are still there, especially in industrial sites. But this also heightens the interest and delight when one has the ability to go to these places and explore them (Interview 65). Traceurs have the means to actualize the affordances of industrial spaces instead of seeing them as hindrances for movement.

Sometimes playing with risks can take dangerous forms, as recalled by one traceur:

> We got into this factory and explored it and there was a smokestack [. . .] There were steps at the side of it and [we thought] that we should go and climb a bit. Well, then we did. Well higher up there was kind of a rack around but it was rather sparse. So, it is interesting how fast you get lactic acid in your muscles even though you have been climbing a lot [. . .] There were three of us and the one who went first reported what was there and there could be some plaster falling and the surface of the steps was rather rusty [. . .] I think I was second and the guy above me started to swear. One of the bars was loose from the other side [. . .] We still moved past it but then it started to be so loose that we decided that we won't go further up [. . .] Later there was surveillance. I know at least one traceur who was brought down from the smokestack by police.
>
> Interview 65

Apart from sound and lighting, the way the space is structured is also important in creating the element of excitement and danger in industrial areas. A rescue dog trainer who we interviewed described industrial spaces as wide and complex: one geocacher described them as 'labyrinthine mazes', with different stages and physical layers of construction creating material complexity (Interview 64; also Anonymous 1). Mapping these dangers is part of the rescue dog training (Figure 8.2). The dogs learn to act in different risk environments and trainers learn how to safely proceed (Interview 64). One airsoft player described 'the three-dimensionality' of the space at old industrial sites as an element of danger. When restricting the area for the game, they had at one point forgotten the possibility that some players might move upwards, resulting in a player shooting others from a dangerous elevated position (Interview 62). The element of risk and danger of playing in these unstable and unfamiliar structures can be seen as an affordance (Hudson, 2014: 303).

What makes these abandoned places good for different activities is, according to the interviewees, a mixture of fascination and the feeling of excitement and danger. For them the dangerous nature of industrial spaces is an affordance for a heightened level of bodily awareness, adrenaline rushes and a greater level of risk-awareness. Using these spaces for different disciplines gives you confidence in your own abilities through training to control your own body. Usually, it is the architecture of these places that enable these different activities to happen. Vast, empty and mentally challenging spaces, big machines, staircases and dark corners of these sites that are not originally designed for sports or other activities,

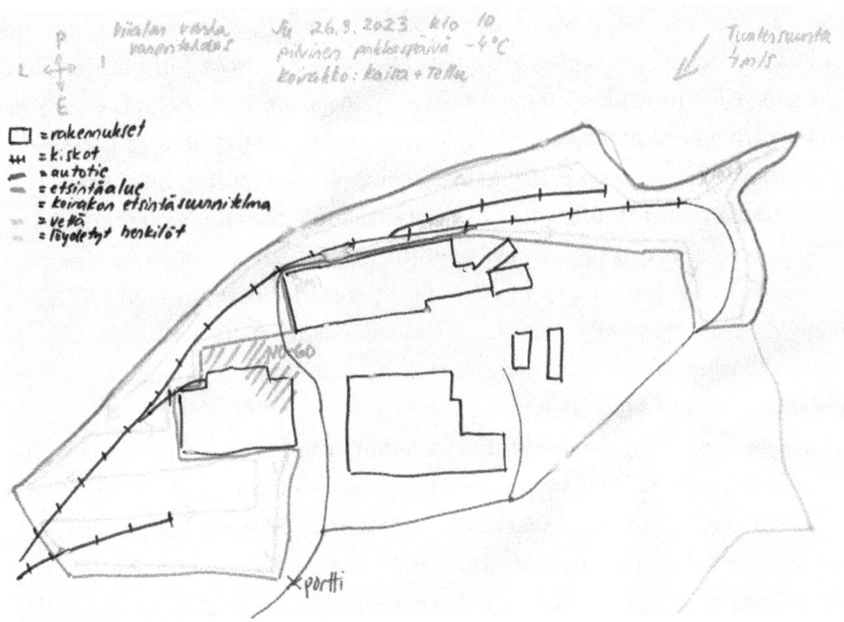

Figure 8.2 A map showing the industrial area marking the area of rescue, the rescued persons and the rescue plan (line with arrows), as well as an area of danger (NO-GO). The map was drawn during the rescue practice (Photo: Interview 64, with permission).

give you the desire for an adventure and an opportunity to let your imagination gallop, was also mentioned by the interviewees.

Even though seeking excitement seems to be a general but often somewhat concealed aspect of many of the activities at industrial sites, one of the interviewees emphasized that one should not go to these places in search of an adrenaline trip, but to have a more respectful attitude and admiration for the place (Interview 60).

Beauty even in ugliness

> A place where nature has taken over can have a charm that is also beautiful.
>
> Interview 60

Admiration for old industrial sites is evident in the ways that our interviewees described ruination as photogenic. Taking and sharing photos has been seen as important, especially for UE (Matila et al., forthcoming) but also other hobbyists, such as traceurs, often share their videos. One urban explorer states that industrial sites are good places for her hobby and a source of inspiration, since 'there is a lot to explore, they are visually inspiring and the spaces are big' (Bunnyurbex). Another urban explorer noted that the impressive size of industrial remains makes them interesting for photographing (Interview 34).

Apart from the magnificent size of the structures, the visible marks of change at industrial sites can be seen as beautiful. One of our interviewees described how seeing trees growing in a place where people formerly worked was mind-boggling (Interview 60). But there was also sadness in witnessing change, especially when coming late to a site after its demolition, or when, for example, the smokestack had just been brought down (Interview 60).

Seeing a space through the eyes of a hobbyist might also turn elements that are generally considered ugly into something beautiful, as they come to be understood as affordances. Ameel and Tani (2012a: 164) note in connection to traceurs how they see concrete as an ideal material and hence value suburban landscapes that can often be seen as boring and ugly. At the training space of the Helsinki Parkour Academy in Vallila, rustic beauty is imitated in the newly built structures (Figure 8.3). One traceur from another city describes the atmosphere in Vallila:

> The light there is awesome because of the big windows with ancient glass and the skylights above. [. . .] And for the equipment they took the same red coloring as in the red girder, a gantry crane. When you go there, you get a great feeling that you are in an old property, industrial property, that has been taken into a totally new use.
>
> Interview 65

The embodied experience of abandoned spaces is a multi-sensory one, and these aspects can become linked to a sense of adventure: 'I would say that the point of going there in the first place, was the atmosphere there. It is the most significant thing. Smells, sounds, and of course everything visual, all that. It was always a small adventure to go to places' (Interview 59). Smell can also be an affordance of the space, as in rescue dog training, where the ways that smells eddy from small cracks or behave in different weather are a part of the dog's experience. This is particularly notable in industrial sites that are left cold or can still have strongly smelling substances (Interview 64). An urban explorer describes how one gets used to odours that most people would not like, for example, the smell of mould, dirt and grease, of old houses or cellars (Interview 59). Another urban explorer notes the positive connotations

Figure 8.3 The Vallila parkour training area incorporates elements from its industrial use and uses these visual elements also in the new structures (Photo: Tiina Äikäs, 2023).

she has with the smell of mould: she also describes her experiences in old houses as powerful, holistic sensory experiences combining sounds, smells and visuality (Interview 60).

Being in abandoned places can hence be a holistic experience. These places can be impressive, merciless and cold, and because of the excitement one gets from the place, one has to even be ready to put up with strong and foul smells. Smells can 'confirm the materiality of space' (Drobnick, 2002: 37) as many of the odours can be connected to the state of ruination of the site as stated above. The abandoned sites can heighten your senses and make you more acutely aware of the surroundings, sounds, smells, surfaces and your reciprocal bodily engagements with them.

Usually, these places have a very unique soundscape as well. The interviewees stated that you can hear water dripping somewhere, wind blowing through the cracks in the walls and the floor cracking under your feet. They also mentioned that these sounds can give a feeling of stagnation of the place, a place where the sound of the built environment and nature meet. Changing seasons can affect the feelings and emotions you get from these places, too. One of the interviewees mentioned that on a hot, summer's day, an abandoned factory can offer a cool place to visit (both physically and mentally) (Interview 60). According to Edensor, using different senses 'is thus central to understanding the interplay between people and space' (Edensor, 2005: 84).

The interviewees problematized the strong emphasis on visuals, which is evident especially in urban exploration, where taking and sharing photos lies at the hearth of the hobby. One said that he never visits a site just for the visual values but is more interested in what has been done there and how and why it is no longer in use (Interview 59). Another stated that there seems to even be some kind of hierarchy among the urban explorers based on who has the best equipment for photo shooting and the most followers for their photos on social media. According to her the most valued explorer nowadays is no longer the one who has visited the most places and has the greatest burn for exploring, but the one who is aesthetically most admired (Interview 60).

In their studies on the way traceurs use the city space, Ameel and Tani (2012a: 169) have noted that: 'Often, environments and materials which are usually interpreted as boring, banal and even ugly are appreciated among traceurs as ideal environments for parkour', and hence they see 'the aesthetics and emotional potential of banal spaces' (Ameel and Tani, 2012a: 171). This is also evident from the way our interviewee describes the use of an old factory site for a yearly parkour event, Supreme Armageddon: 'There are really fine rail structures and [. . .] concrete blocks that have worked as bases for paper machines. And different stairs and pipes from which you can hang, and they hold it and all. And then there are just holes on the floor, which are interesting to climb or jump over. [. . .] We could have workshops there in an authentic, abandoned factory' (Interview 65). Here a space considered hazardous for most people is seen full of affordances for practicing.

Spaces of imagination and playfulness

> Actually, always when you go to a place like that, it is like watching a really good movie or reading a really good book, which takes you away from this moment and real life and you step into another reality. You go there inside other inner walls and there is another world.
>
> Interview 59

Our interviewee stated that this sense of otherworldliness was what fascinated him the most in urban exploration. Abandoned places inspire the imagination, they make us wonder what happened to the people who worked and dwelled within these walls. This can be emotionally laden when witnessing the traces of past lives, such as children's shoes, as one of the interviewees noted (Interview 60). This kind of intimately personal material culture in particular can let imaginations run wild.

In some cases, the feeling of fear born from abandoned material culture, dark corners and unusual soundscapes can be heightened when it chimes with horror culture aesthetics. Smelling the mould, dirt and blood-like rust can also fuel these evocations of horror film locations. Researchers have noted the similarities between UE imagery and horror films (Bennett, 2011: 426 referring to Wilson, 2003), which are evident, for example, in the photographs taken by one of our interviewees with the pseudonym Bunnyurbex. She is purposefully flirting with horror and fetish imagery in her photography of abandoned sites that she explores. The aesthetics of abandoned industrial sites is a significant dimension of what has come to be called 'ruin porn': photography focusing on the perceived melancholy and romantic aesthetics of deindustrialization and its wider social and material processes (e.g. Lyons, 2018). Ruin porn is a powerful and pervasive cultural strand in social media, such as Instagram, but also in popular publications and other media. In response, some critics have highlighted the voyeuristic and sentimentalizing functions of ruin porn, and the elitism of middle-class explorers and photographers claiming ownership of working-class heritage (Strangleman, 2013).

Bunnyurbex describes how she often explores sites together with her mother, who also acts as her photographer. In fact, many of the hobbies that we explore here, despite their postindustrial settings, offer chances for families and different generations to go exploring together. Geocaching allows the safest and most curated way to do this, including with children, but also urban exploring, airsoft and paintball war games allow different generations to come together.

Geocaching is at its heart a playful and corporeal engagement with your surroundings in a virtual treasure hunt, where one tries to grasp the unseen world(s) of caches often hidden in plain sight, beyond the veil of everyday normality. This is described among the hobbyists as 'something that the muggles won't see' ('jästi' in Finnish), referring to the naming of non-magical persons in the Harry Potter novels. 'Muggling' ('jästiä' in Finnish) is a term used by the geocachers to describe a situation where a cache has been revealed to non-geocachers and is in danger of being destroyed. This engagement with popular culture is also evident in the advert for the Supreme Parkour Armageddon in 2022, where a reference is made to the Middle-earth from J. R. R. Tolkien's tales by describing the event as 'The escape from Middle-Örth'. These kinds of imaginative references serve to connect practitioners to alternative worlds.

Naturally, not all such associations are positive. As mentioned above, the atmosphere of old industrial sites can resemble a site of war, bringing to mind the recent images from Ukraine and elsewhere. This can be used not only in urban survivalism but in airsoft games, where the warlike feeling is enhanced by guns and military-like clothing. In the studies of paintball practitioners, a game similar to airsoft, Hanemaayer notes that 'the game [paintball] is a site where imaginative war scenarios illuminate the desire for enjoyment, play and role-taking'. There is a possibility of risk seeking and playing with death without actual risk, and an interplay between the real and the pretend, e.g. the use of paint instead of lead bullets. (Hanemaayer, 2011: 235, 241).

These make-believe and imaginative worlds can be seen as marks of playfulness, often associated especially with parkour (Geyh, 2006; Stevens, 2007; Ameel and Tani, 2012b: 22; Larsen, 2021: 26, even though alternative ideas have also been explored by Aggerholm and Larsen, 2017). The way parkour can transform even the most alienated environment into a place of play has been considered something very human (Larsen, 2021: 26), one could say especially for *Homo Ludens*. Playfulness can be seen as a way to take a fresh perspective on one's environment and to test and expand its limits and affordances (Ameel and Tani, 2012b: 21–2, see also Stevens, 2007). Parkour has been described as 'perceiving beyond the discursively constructed meanings of objects and spaces and finding personal, creative ways to use the space' (Ameel and Tani, 2012a: 165).

Both of our parkour interviews mentioned the playful nature of the practice: 'Traceurs are often playful and adventurous by nature so the abandoned factory created a good sense of adventure for the practice. [. . .] The idea of an abandoned factory, which is not created for our purposes but we can ourselves go there to explore and interpret the place by our own means' (Interview 63). The 'childlike inventiveness' (Ameel and Tani, 2012b: 18) is evident in the way that traceurs see the environment. The same interviewee continues: 'We don't need anything special to get the practice started, we just have to use our creativity and imagination. You can spend the whole day jumping on a couple of natural stones or paving or over a fence' (Interview 63). Here what others see as barriers in the environment – and which are often intended as such – are transformed into playful obstacles (see also Kidder, 2012: 230).

The notion of abandoned industrial spaces as opportunities for 'childish' creative play harks back to the idea of adventure playgrounds or 'junk playgrounds'. These were pioneered in Denmark and the UK during the Second World War, making use of cleared bombsites and development areas. The aim of founders such as Marie Paneth was to create spaces where troubled or traumatized young people could enjoy free play, exploring and actively re-shaping their material surroundings (Kozlovsky, 2007). Beyond these semi-formalized spaces, young people have long made use of war-damaged buildings as play spaces, enjoying the same calculated risks and freedoms from surveillance that attract hobbyists to abandoned industrial sites (Moshenska, 2014).

To preserve or not to preserve

> Not too much is taken away and put plexiglass in front and all weird virtual stuff and other stuff that doesn't originally belong there.
>
> Interview 59

In a citation in the previous section, the interviewee emphasizes how great it is to use 'an *authentic*, abandoned factory' (our cursive). For many of our interviewees, the authenticity stems from the unrenovated form of the building, where things are left as they were. This state of abandonment offers affordances for uses that were not originally planned for the building. One traceur comments that for them a building is more usable if it is not renovated too much but left 'rough'. It is cheaper and they do not need to worry that their activities will damage or degrade the space. Thus they can reshape the space without being too careful, but also use the original structures, such as bases for machines and lofts. 'You can flirt with the space and the things that are already there' (Interview 65).

There were also contradictory views. One urban explorer notes that when she sees places that are being renovated, her first thought is that it is lost for her. But on the other hand, she likes it when the old is appreciated and repaired. She points out that when these places are turned into visitor sites it slows down their final destruction and vandalism but also changes them. They lose their meaning as abandoned and leftover places not valued by the rest of the society, and the anarchism of valuing the place is turned into something shared by all (Interview 60).

For the interviewees, the desire to protect the sites collides with the perceived boredom and loss of freedom in musealized spaces. One interviewee suggests that it would be good if at least some parts of the old structures were left in their wilder milieu to remind people of the heritage of the site (Interview 65). There is a strong ambivalence in the way many of the practitioners respect and are inspired by the history of the places and want to protect them, but also admire the sublime nature of ruination and overgrowth and its beauty as stated earlier. Restoring a site can at the same time save it and make it not usable for the practitioners.

Conclusion: derelict and in use

> There reverberates a bit this kind of a weird echo of a post-industrial era, that there have really been all train rails and all systems here and now this is just standing here empty. It's also quite impressive, in addition to being ruthless.
>
> Interview 61

As we have shown, the hobbyists see old industrial sites mainly through the things they afford for their practices, but also as heritage and reminders of the industrial past. If we return to geocaching, of which we have our own personal experiences, it has been noted that geocaching shapes the user's experience not only of space but also of cultural heritage (e.g. Buccitelli, 2017). Geocaching can be seen to have formed a special relationship with heritage. In Finland there is a clear statistical correlation that shows how geocaches have tended to be placed at or close to heritage sites (DHH17, 2017; Immonen, 2022: 50). Seeking a geocache might give an added excitement to visiting a heritage site and vice versa. But as noted above, for other hobbyists the heritagization and musealization of old industrial sites poses a challenge, restricting their use and access.

At the heart of many of the activities discussed above there is the ability to see the potential in places deemed abandoned, and to use them in innovative ways beyond their original intent. One interviewee even describes his irritation when he is told that, 'You are not supposed to climb there – you are not allowed to climb there. It's not built for that.' He rightly notes that,

> People are not usually doing with objects only the things that they are meant for. Is a ball made for kicking or is it made for throwing or for rolling? What is it made for? What is it that you are supposed to do or not supposed to do? In that way, it is a question of how we use and take possession of the urban environment, and I think it is good that the traceurs challenge it.
>
> Interview 65

Similarly, Kidder (2012: 237) states that 'traceurs are offering a lived critique of the built environment – asserting playful creativity in the dead spaces of the city'.

Abandoned industrial sites are often considered wasteland, or to use Marc Augé's ([1992] 1995) concept 'non-places', places of transience, which humans only pass by or visit temporarily and which do not hold enough meanings and significance to become places. Larsen (2021: 30) has stated that the number of these kinds of places where the link between past and present as well as between people and environment is missing, grows as we stop paying attention to our surroundings. In contrast, Hilary Orange (2016: 208) has explored the 'reanimation' of such spaces: 'the ways in which industrial landscape continues to be lived in, reacted against and claimed.' For Orange (2016: 15), heritage research can be a part of the reanimation process: the everyday experiences of contemporary explorers woven together with the memories of former factory workers, and the ghosts of history that haunt these spaces.

Our interviewees demonstrated their multisensorial, bodily awarenesses of their environments and of the affordances they offer. For them, the affordance is an interplay between personal skills, bodily engagement, architecture and atmosphere that affects how a person can use a space. Affordances are dependent on the way one looks at the world; what one might see as a site to be demolished, another can see as a place of adventures (cf. Rietvelt and Kiverstein, 2014: 327). Rietveld and Kiverstein (2014: 326), when talking about animals, refer to affordance as 'the exercise of an ability in a specific context'. This ability can also refer in our cases to an ability to see something in the environment that others do not or cannot (also Larsen, 2016: 301; 2021: 36; De Martini Ugolotti and Genova, 2023). In the context of parkour, this ability has been described as 'parkour vision' (Kidder, 2012: 245) or 'parkour eyes' (Ameel and Tani, 2012a), through which one can re-imagine what the environment can afford. The same goes with the geocachers 'magical vision', as described by one of our informants, the 'geocacher's sight of a hidden world' with endless hiding places and cache potentials it affords that the muggles or non-initiates will not grasp (Anonymous 1).

The vision of the hobbyist turns spaces that are traditionally approached with negative emotions or as marginal spaces into practical and suitable terrains, as well as aesthetic and emotional places, as they are connected to valued practices (see also Ameel and Tani, 2012a: 168; Hudson, 2014: 194). They make playful use of spaces that for others seem to lack meaning and interest (Kidder, 2012: 237). As Ameel and Tani (2012a: 170, see also Relph [1976] 1986) have put it:

> 'Parkour eyes' [or 'geocacher's sight' (Anonymous 1)] are not only about seeing possibilities in unexpected places, but also about seeing possibilities for attaching new and unexpected feelings to a site, thus transforming mere, dead space into a meaningful place. Seen with such eyes, spaces that had been largely detached from links to history or location become inscribed again with an appreciation of their symbolic significances and their identity. By developing an embodied relation with largely inauthentic environments or non-places, traceurs gradually create, by way of their embodied activities, what amounts to a more authentic sense of place.

Also, geocachers engage often bodily and intimately with their surroundings, running their fingers under railings and windowsills, crawling into narrow spaces and behind flower bushes, or opening manhole covers and hidden hatches in search of their prize: the cache and the log book where they can scribble their alias.

Kidder (2012: 247) notes that 'through parkour, individuals re-imagine the city. The abstractions of capitalism can be turned into something joyfully lived – in ways never intended by their designers'. This re-imagining and new ways of using space is a recurring theme in many of our interviews. It involves revealing and opposing capitalist ideas of the uses of urban space, and resistance to alienating and restricting approaches to planning (Daskalaki et al., 2008; Larsen, 2021: 26). Critique of capitalist commodification of urban spaces can also be intertwined with a more personal bodily relation to the environment. In our interviews both of these are evident (see also Ameel and Tani, 2012b: 23; Garrett, 2014).

Despite their sometimes anti-authoritarian attitudes, many hobbyists share a principle of leaving the site as they found it (e.g. Interview 65, also Lamb, 2014: 115). Airsoft practitioners even told us how they clean the site after their game. There is a fine balance here, in which a site is not unnecessarily disturbed either by damaging it or by renovating it. The affordances of industrial sites for hobbyists can be lost both through their demolition and through their musealization. This ambivalence is also evident in the quote from our interview:

'Well, there is a certain contradiction. In a way, when you sometimes see places that are started to be renovated, you get the feeling that damned now that one is gone too and I didn't have time to visit it. But then you also feel that it is nicer if old is appreciated and fixed than if it is demolished and some glass palace built at its place.'

Interview 60

References

Archives

Smokestack Memories interviews will be archived to the Labour Museum Werstas in Tampere, Finland.

Anonymous 1–2, personal archives of Oula Seitsonen.

Literature

Aggerholm, K. and S. H. Larsen (2017), 'Parkour as acrobatics: an existential phenomenological study of movement in parkour', *Qualitative Research in Sport, Exercise and Health*, 9 (1): 69–86. doi:10.1080/2159676X.2016.1196387.

Ameel, L. and S. Tani (2012a), 'Everyday aesthetics in action: Parkour eyes and the beauty of concrete walls', *Emotion, Space and Society*, 5 (3): 164–73.

Ameel, L. and S. Tani (2012b), 'Parkour: Creating loose spaces?', *Geografiska Annaler: Series B, Human Geography*, 94 (1): 17–30. doi:10.1111/j.1468-0467.2012.00393.x.

Augé, M. ([1992]1995), *Non-places: Introduction to an Anthropology of Supermodernity*, trans. J. Howe, London: Verso.

Bennett, L. (2011), 'Bunkerology – A Case Study in the Theory and Practice of Urban Exploration', *Environment and Planning D: Society and Space* 2011, vol. 29: 421–34.

Buccitelli, A. B. (2017), 'Locative Gaming, Folk Geographies, and the Experience of Cultural Heritage', *Cultural Analysis*, 16 (1): 7–29.

Daskalaki, M., A. Stara and M. Imas (2008), 'The "Parkour Organization": Inhabitation of Corporate Spaces', *Culture and Organization*, 14 (1): 49–64. doi:10.1080/14759550701659029.

De Martini Ugolotti, N. and C. Genova (2023), 'Parkour, Graffiti, and the Politics of (In)Visibility in Aestheticized Cityscapes', *Space and Culture*, 0, no. 0. https://doi.org/10.1177/12063312231155356.

DHH17 (2017), Helsinki Digital Humanities Hackathon 2017. https://www.helsinki.fi/en/helsinki-centre-for-digital-humanities/dhh-hackathon/helsinki-digital-humanities-hackathon-2017-dhh17.

Drobnick, J. (2002), 'Toposmia: Art, Scent, and Interrogations of Spatiality', *Angelaki: Journal of Theoretical Humanities*, 7 (1): 31–47. doi:10.1080/09697250220142047.

Edensor, T. (2005), *Industrial Ruins, Space, Aesthetics and Materiality*, Oxford and New York: Berg.

Edensor, T., B. Evans, J. Holloway, S. Millington and J. Binnie (2011), 'Playing in Industrial Ruins: Interrogating Teleological Understandings of Play in Spaces of Material Alterity and Low Surveillance', in A. Jorgensen and R. Keenan (eds), *Urban Wildscapes*, 65–79, London: Routledge.

Garrett, B. L. (2014), 'Undertaking recreational trespass: Urban exploration and infiltration', *Transactions of the Institute of British Geographers*, 39 (1): 1–13.

Geyh, P. (2006), 'Urban Free Flow: A Poetics of Parkour', *M/C Journal*, 9 (3): no pages. https://doi.org/10.5204/mcj.2635.

Gibson, J. J. ([1979] 1986), *The Ecological Approach to Visual Perception*, Hillsdale, NJ: Erlbaum.

Hanemaayer, A. (2011), 'Game death: Fantasy and mastery in scenario paintball', in T. Connolly (ed.), *Spectacular Death: Interdisciplinary Perspectives on Mortality and (Un)representability*, 233–48, Bristol and Chicago: Intellect.

Heft, H. (2010), 'Affordances and the perception of landscape: An inquiry into environmental perception and aesthetics', in C. Ward Thompson, P. Aspinall and S. Bell (eds), *Innovative Approaches to Researching Landscape and Health*, Open Space: People Space 2: 9–32, New York: Routledge.

Hudson, J. (2014), 'The affordances and potentialities of derelict urban spaces', in B. Olsen and Þ. Pétursdóttir (eds), *Ruin Memories*, 193–214, New York: Routledge.

Huizinga, J. (1938), *Homo ludens: Proeve eener bepaling van het spel-element der cultuur*, Haarlem: H.D. Tjeenk Willink.

Immonen, V. (2022), 'What is Queer Heritage? Queercache and the Epistemology of the Closet', *Lambda Nordica*, 27 (2): 41–68.

Kidder, J. L. (2012), 'Parkour, The Affective Appropriation of Urban Space, and the Real/Virtual Dialectic', *City & Community*, 11 (3): 229–53. https://doi.org/10.1111/j.1540-6040.2012.01406.x.

Kozlovsky, R. (2007), 'Adventure Playgrounds and Postwar Reconstruction', in M. Gutman and N. de Coninck-Smith (eds), *Designing Modern Childhoods: History, Space, and the Material Culture of Children; An International Reader*, 171–90, New Brunswick: Rutgers University Press.

Kyttä, M. (2004), 'The extent of children's independent mobility and the number of actualized affordances as criteria for child-friendly environments', *Journal of Environmental Psychology*, 24 (2): 179–98.

Lamb, M. D. (2014), 'Misuse of the Monument: The art of parkour and the discursive limits of a disciplinary architecture', *Journal of Urban Cultural Studies*, 1 (1): 107–26.

Larsen, S. H. (2016), 'What can the parkour craftsmen tell us about bodily expertise and skilled movement?', *Sport, Ethics and Philosophy*, 10 (3): 295–309. doi:10.1080/17511321.2016.1217919.

Larsen, S. H. (2021), 'Parkour: Playing the modern, accelerated city', *Journal of the Philosophy of Sport*, 48 (1): 26–44. doi:10.1080/00948705.2020.1834862.

Lyons, S. (ed.) (2018), *Ruin Porn and the Obsession with Decay,* Cham: Palgrave Macmillan.
Matila, T., T. Äikäs, O. Seitsonen, E. Koskinen-Koivisto and V.-P. Herva (forthcoming), 'Smokestack nostalgia or hypermodern spirituality? Urban exploration in abandoned factories'.
Moshenska, G. (2014), 'Children in ruins: Bombsites as playgrounds in Second World War Britain', in B. Olsen and Þ. Pétursdóttir (eds), *Ruin Memories: Materiality, Aesthetics and the Archaeology of the Recent Past*, 230–49, Abingdon: Routledge.
Ninjalicious (2005), *Access All Areas: A User's Guide to the Art of Urban Exploration*, Toronto: Infilpress.
Orange, H. (ed.) (2016), *Reanimating Industrial Spaces: Conducting Memory Work in Post-Industrial Societies*, London: Routledge.
Raymond C. M., M. Kyttä and R. Stedman (2017), 'Sense of place, fast and slow: The potential contributions of affordance theory to sense of place', *Frontiers in Psychology,* 8 (1674): 1–14.
Relph, E. ([1976] 1986), *Place and Placelessness*, Repr. London: Pion.
Rietveld, E. and J. Kiverstein (2014), 'A rich landscape of affordances', *Ecological Psychology,* 26 (4): 325–52. doi:10.1080/10407413.2014.958035.
Stevens, Q. (2007), *The Ludic City: Exploring the Potential of Public Spaces*, London and New York: Routledge.
Strangleman, T. (2013), '"Smokestack Nostalgia," "Ruin Porn" or Working-Class Obituary: The Role and Meaning of Deindustrial Representation', *International Labor and Working-Class History,* 84: 23–37. https://doi.org/10.1017/S0147547913000239.
Wilson, H. (2003), 'Pipedown', *The Guardian*, 5 July 2003.

9

Remaining without Preservation: The Zombie Standing of Kino Rodina in Estonia

Francisco Martínez

Introduction

What to preserve and how to do so are key challenges of the present. As heritage is *past socially made durable*, contingent elements of the present have to be cut off. Then, they are presented as something that has to be sacrificed for clarity and fixity. The question is how, and by whom, and the answer has to be found in situated, empirical cases. For instance, there is a building in Sillamäe that has been listed as heritage for twenty-three years; however, the resources to fix its forms and reduce its material instability have not been found and mobilized since.

What you see is what you get. And here you only see matter falling apart, grass growing on the ground, and broken glass in that corner, and that other, and this one. For a building to be considered as ambivalent heritage, its fate as a concrete object has to under question, while its singular value remains institutionally recognized. Perhaps, after all, ambivalent heritage can be everything plus institutional valuation, only material care. The noun ambivalence refers to having mixed feelings and facing multiple directions at once, thus to a difficult fitting in the present. According to Zygmunt Bauman (1991), this phenomenon is the side effect of the ordering obsession of modernity. Ambivalent heritage therefore is intrinsically contemporary, mirrors the social relations around and projects us as a collective.

In this chapter, I will examine a case of postindustrial ambivalent heritage that is not a factory, but a cinema built for its workers. Industrial activity, closely connected with modern infrastructural ideas, often came along with venues such as cinemas, theatres, sport halls, military bases, hospitals, schools and media offices. In Sillamäe, some of them stand as ruins, such as *Kazarma*, a dorm of the Soviet army; or the building on Rumjantsevi tn. 3, once the headquarters of the Sillamäe Herald (*Силламяэский Вестник*) newspaper. Some others, such as the central school on Mere pst. 3, have been repurposed into a private cultural hub, which plans to host a restaurant, an art gallery, offices, a beauty parlour, and apartments. An important question, thus, is ownership: an aspect that will help us understand a key difference between the two buildings listed as heritage in Sillamäe – the Cultural Centre and Kino Rodina (Motherland Cinema).

The learnings of this chapter are thus twofold. Firstly, that industrial heritage can also be found in sites that are related to factories only indirectly, such as electrical grids, roads and ports, but also schools, hospitals and leisure facilities, such as cinemas. As we see in the case of Sillamäe, modern industrial activity was intricated within infrastructure-making and the process of deciding what kind of ideas and substances we wish to pass on. And cultural infrastructure played a role in that. Secondly, we can define ambivalent heritage as everything that failed to go away and remains against the grain, creating discomfort in those around. Indeed, a sense of annoyance, and even embarrassment, is associated with the building of Kino Rodina, one that is not necessarily related to its original function, or how the building was constructed (mostly by German and Estonian war prisoners), but because of the actual unfitness in the present.

Almost heritage

What can be the archaeological evidence left behind in a settlement where uranium was industrially processed, such as Sillamäe? And how can we accommodate what does not fit in the present? In the eighteenth century, the northeastern area on the Baltic Sea was populated by fishermen and later on it became a popular holiday resort among Russian intellectuals. However, after the construction of a uranium enrichment plant, the rather peaceful site was turned into a closed town, which could only be accessed with special permits.

Sillamäe was developed from ruins as an industrial and military town supplying nuclear materials to energy plants and weapon facilities until 1989. During that period, the settlement was given the code name 'R-6685', and thousands of workers from across the Soviet Union were relocated to a place that did not appear on the official maps. Intriguingly, we can say that Sillamäe was always there, but it never appeared as a town. In the late 1980s, it had a population of more than 20,000 people; nowadays it is a mere 12,000. The Sillamäe plant was privatized and renamed as Silmet in 1997. It belonged to the US mining group Molycorp until 2016, when it filed for bankrupt. Then it was acquired by the

Canadian Neo Performance Materials to process rare minerals that are later sold in the global electronics markets.

Sillamäe's architectural core is characterized by a neoclassical, Stalinistic style, crowned by a monumental promenade, which has a House of Culture and a municipal building on the top of the hill. Kino Rodina is not far away, just a few metres further west. These buildings were built in the middle of the twentieth century by war prisoners and military personnel, young men who were mobilized into the Estonian and German forces. 'Builders were needed and there were no professionals. The local leaders made a proposal to the prisoners who lived here at the centre, under the nowadays' promenade. They promised to reduce the detention time for those who were willing to work. So, to some extent, we can say that they were volunteers though only 2% of the workers were really free', explained Aleksandr Popolitov at the Sillamäe Museum (which was located in the bomb shelter of the House of Culture until 2014).

The House of Culture has 400 seats and the cinema 612 seats; thus, the entertainment of local workers was well provided for. The monumentality of these buildings mirrors the official understanding of culture by the Soviet power, exhibiting a totalitarian control over both ideas and the landscape (resonating along the modern assumptions of a clear-cut separation between culture and nature).

For more than a decade, dedicated officials of the Estonian Heritage Board have been trying to include over 150 buildings in the city centre under protection. Over the years, multiple reports have been written for that purpose;[1] but, three kinds of problems have made their effort still unsuccessful: (1) Underfunding of the Heritage Board, which means overwork for the officials and limited financial support for the owners; (2) Opposition among some local neighbours, as the listing as heritage protection would mean for them a limitation to refurbish apartments as they please, as well as an increase in the costs; (3) Overly negative national takes on Soviet legacies and a troublesome geopolitical context. In other words, Sillamäe is widely labelled as a 'Russian' town in the rest of the country; the Estonian state rebuilt itself in opposition to its Soviet past and the imposed Russification; and the war in Ukraine has generated divisions and tensions within society.

Nonetheless, and as written above, two buildings in Sillamäe were listed as heritage in 2001. In the early 1990s, a discussion was held to discuss what to do with these buildings. The House of Culture remained in public hands and, a decade ago, it underwent renovations financially supported by the European Union. The result is a bit Napoleonic, overdone, but its condition as heritage is

Figure 9.1 The House of Culture, lobby and basement (Photo: Francisco Martínez, 2022).

Figure 9.2 The House of Culture, lobby and basement (Photo: Francisco Martínez, 2022).

secured. The House of Culture was designed by Aleksandr Popov in 1946–7 and built in 1947–50. In the basement, we can still visit an exhibition foregrounding, in a nostalgic way, the industrial origins of the actual Sillamäe, including Soviet flags and busts of Communist leaders.[2]

However, Kino Rodina fell into private hands. It was bought during the Perestroika years and functioned as a cinema until 1 October 1995. Then, one of the halls, the smaller one, was used as a nightclub for several years. The club was named 'Korrida', and neon green and pink colours were added to the neoclassical ornaments and walls to create a more festive atmosphere. Aleksandr Starodubtsev, one of the three owners of Kino Rodina, states during our visit that he decided to close the club when drugs started to circulate among clients, even though Korrida was very popular among local residents. Aleksandr repeats that he is old and lacks the motivation to do anything with the building. He also complains about the electricity bill being too expensive. All in all, the owners are unable to fix the building and sell it for 200,000 Euros.[3]

I have been lucky enough to finally visit Kino Rodina. I had four unsuccessful attempts to enter the building. One evening I even tried to trespass into Kino Rodina illegally; however, the doors and windows were well sealed, and the only entry point, a sort of hole on the second floor, was too risky. Finally, Allen Allet, an architect who acts as a head of local urban planning, managed to arrange the visit. He had been working in the municipality of Sillamäe for more than two years and his attempts to visit the building were also unsuccessful until August 2023. Apparently, communication with Aleksandr was not always fluid.

And here we are, inside this ambivalent heritage, our Moby Dick, and in the middle of a highly sensorial and reflective experience. We all meet in front of the impressive façade decorated with figured moulding; it also has a sign that reads 'Kino Rodina' in a beautiful old school script. Albeit in a degraded condition, the building is still a key part of the cityscape, and also a signifier of progress and prosperity symbolized by modern infrastructure. Aleksandr rides a bike and Allen comes walking. He has also invited Vladimir Gruzdev, owner of the new cultural hub at the former school, and I have invited along photographer Riina Varol.

For forty minutes we stroll around the cinema carcass. Upon entry, Aleksandr guides us, at first, and then rather laconically walks with us. At some point, he gets bored and goes back to the entrance. Now, I can step into whatever intestine I like, but merely for fifteen minutes more. When I came back to the lobby to check what the situation was, I saw Aleksandr standing at the door with the keys in his hand. Then, he asked if the photographer had gotten lost, so I understood

Figure 9.3 Current state of Kino Rodina, as at August 2023. By Riina Varol.

Figure 9.4 Current state of Kino Rodina, as at August 2023. By Riina Varol.

Figure 9.5 Current state of Kino Rodina, as at August 2023. By Riina Varol.

Figure 9.6 Current state of Kino Rodina, as at August 2023. By Riina Varol.

Figure 9.7 Current state of Kino Rodina, as at August 2023. By Riina Varol.

Figure 9.8 Current state of Kino Rodina, as at August 2023. By Riina Varol.

that it is the moment to leave. Probably, I will not ever have the chance to enter Kino Rodina again, but when going away I am afraid of looking back as if I would be turned into stone, as with Medusa and with Orpheus. After all, I was coming out from the underworld.

The building has an L-shaped ground plan with two halls and two floors. It shows a rich, ornamental decor both in the interior and exterior design. The front room of the cinema was a spacious lobby, and on the second floor there was a restaurant with a dance hall. Nowadays, the chairs are laying on the floor, a short glass for vodka is hidden in the atrium of the waiter, and a package of chocolate with fungi is forgotten in a corner of the kitchen. I am surprised there are so few tape reels around, and none in the projection room. On the contrary, there are several wigs and garlands around, as if someone had celebrated New Year's Eve in here. Perhaps in 1997, or in 2004 or in 2013.

The architectural project was designed by the architectural office 'Giprokino', the Soviet State Institute for the Design of Cinema Theatres across the empire. Indeed, a similar movie theatre was built in Ashgabat, Turkmenistan. The construction work began in 1952 and the grand opening took place on 1 May 1955, with a huge gathering of people. The façade and the interior were designed by the architects Stenyushin and Kurennoy. The light-yellow colour of the walls combines well with white mouldings above the doors and windows and under the roof canopy, giving a sense of solemnity to the building.

Crystal chandeliers hang in the lobby and auditoriums. At the angles of the lobby, to the right and to left, there are two wide-screen cinema halls (a neon pink colour from the time of the Korrida club asks for attention in one of the halls). On the second floor of the building, there was a restaurant with a dance hall. There used to be a billiard room and a shooting gallery in the basement too, though I did not see any traces of them. Indeed, the door that unlocks the basement was the only one I could not open or force, precisely the one that keeps on lingering in my mind.

Alfidina Orlova worked in the city library from 1960 to 2004. She has written several books about the cultural history of Sillamäe, for the new generations to know how it was, and for the old-timers to remember the beginnings, as she explains. In the book *Sillamäe 1946–1989*, Orlova (2005) divides the history of the settlement into two periods: before and after the Perestroika. In her view, the first half (characterized by the industrial processing of uranium) can be also divided into two periods, the village of Sillamäe (from 1946 to 1956), and the city of Sillamäe (from 1957 to 1991). So, local history begins in 1946 for Orlova. However, industrial activity had started already in 1928, when an oil shale processing plant and a power plant were opened by the Swedish *Estländska oljeskifferkonsortiet*. And the first mention of a settlement in this area dates back to 1502, referencing a tavern for pirates un by someone called 'Tog-Brugen'. Back then, the territory belonged to the Livonian order.

In one of the passages included in Orlova's book (2005: 135–7), she recollects that there was a buffet in the lobby of Kino Rodina, and between sessions a brass band played and amateur singers from the House of Culture would sing there. Tarzan was the first Western film shown in Kino Rodina, and every day there were six sessions. Nowadays, though, the building stands in a precarious material equilibrium and there is no sign of Tarzan. In turn we find peeling-off ceilings, invoices on the floor, dirty curtains, broken windows, empty bottles, calendars from 2008, ashtrays, audio cassettes, bingo tickets, wardrobe tokens, cinema lenses of various colours, handwritten signs, dusty banners, wires and more wires, the working clothes of the waiters, a poster of the war drama 'The Alive and the Dead' (*Живые и мёртвые*, 1964, directed by A. Stolper), and the detailed drawings of the building to be shown to potential real estate investors. That is the actual reality of Kino Rodina. For the current owners it is cheaper to pay the fines sent by the Heritage Board for not taking enough care of the building than to invest in its repair. Hence, what we encounter is the practical absence of material intervention and historical interpretation for more than twenty years.

Between abandonment and recognition

Modern ruins recreate the illusion of a particular bonding that is now missed, so they might potentially anchor a sense of belonging to a community and also of self-worth (Pusca, 2010; Martínez, 2024). Ambivalent industrial heritage is slightly different though: it shows a liminal standing between preservation and recognition, life and death, profit and loss, use and abandonment. Perhaps it is because this kind of heritage is necessarily a landmark of two kinds of cycles: industrialization and deindustrialization, a landmark that has no place in the present, nor in nation building.

Before our visit, Allen warned me 'not to ask too artistic questions', as I was introduced to Aleksandr as a person 'with connections in the real estate business'. In our previous encounters, Allen has told me that he would like the urbanity of Sillamäe to recover, and for that purpose, external investment has to come to rejuvenate key abandoned buildings, such as Kino Rodina, the school in Mere pst. 3 and the former headquarters of the newspaper. Industrial investments around the port are, of course, important, but the depauperated standing of key buildings at the centre affects not just the visual order of the town but also how it is culturally represented and cognitively internalized by locals – thus having epistemological consequences and potentially becoming a target of vandalism. However, the municipality has limited financial capacity to deal with these scars (Storm, 2014), especially in a context of negative capability characterized by outmigration and industrial transition towards a green model.

Buildings are still a repository of public representations and collective bonds despite standing in ruination. They do not simply express pre-existing identities or social relations, but also enable people to become neighbours and citizens in a particular way. However, the anthropological implications of destructive practices in architecture remain understudied (Buchli, 1999; 2013). Kevin Lynch (1990), for instance, argued that architects should plan the demolition of their building at the same time the design is being made. Also, Neil Harris (1999) suggested studying buildings as entities that follow life stages, such as birth/initiation, childhood/coming of age, and finally, death/burial/sainthood. However, buildings do not simply live their own way apart from society; likewise, they are not born, but constructed, maintained or abandoned – always based on all-too-human decisions. In this sense, buildings are difficult to disentangle from present politics, socio-economic processes and the subjectivity of their tenants or users.

The preservation of a building also entails the signification of how we want to define ourselves, as well as orders of worth in society. Buildings are objects where public and private – collective and individual – visions meet, holding together different dimensions of care and responsibility. The death of buildings is ultimately a matter of socio-political choice in the present, drawing attention to both abilities and incapacities of the agents that are supposed to be responsible. Likewise, the deterioration of built forms implies new risks and a reminder of the inevitable fate of architecture: buildings are considered to have relatively high and stable value because of their durability, but as long as they stand, built forms face the ever-present threat of falling apart (Cairns and Jacobs, 2014).

In short, buildings are simultaneously material and social; they are not just built and maintained based on utilitarian reasoning, but they also embody larger political visions and activate other kinds of relations. The question of how strategies and practices of preservation meet the ground is therefore crucial, making tangible and visible the way in which the sustainability of architecture is implicated in the political, and vice versa. Indeed, neglected buildings bear witness to a particular structure of feeling, revealing wider relations between people, and between people and institutions. In the case of Sillamäe, the ruined standing of real estate negatively shapes the ways in which this town is encountered and imagined by both visitors and residents. Some of the residents, indeed, referred to their actual abandonment as a threat to local urbanity.

Nevertheless, there are different ways of measuring the urbanity of a settlement. One is demographic and statistical, such as the Eurostat (2018), which uses a threshold of 5,000 inhabitants and establishes a density threshold of 300 people per square kilometre. However, this method does not take into account the actual use of the existing infrastructure and real estate. Then, in a pioneering definition put forward by sociologist Louis Wirth (1938: 23), urbanity was presented as 'numbers, density and heterogeneity', including physical structures, institutional organizations and a set of attitudes and ideas as the elements making urbanity possible.

Another theory addressing urban deterioration (rather influential among policy makers in the 1980s), was the 'Broken Windows' thesis, inspired by how police officers tried to maintain public order in New Jersey in the 1970s. As James Wilson and George Kelling (1982) argued, order relies on a set of material practices, and our encounter with built forms and façades reinforces messages of care and concern, or vice versa. As they put it, 'If a window in a building is broken and is left unrepaired, all the rest of the windows will soon be broken' (ibid.: 31). Hence, clear evidence of neglect, political abandonment and disrepair should be eradicated because these symptoms are contagious and lead to disorder and detachment.

Wasted legacies

In Estonia, the wasting of Soviet legacies was part of a wider political process of disqualification, enmeshed in new relations of power. In their devalued traces, contemporary wasted legacies appear as a tangible, forensic manifestation of an uneasy integration of the Soviet past into the present. To be categorized as wasteful implies that one no longer has a purpose or worth. Also, the very wasting of a legacy reinforces a sense of residuality and makes sensorially perceptible the discourses attributing to socialist ideas an 'intrinsic failure'. The post-socialist change of political and cultural scripts followed a logic of sacrifice, making it 'necessary' to abandon certain meaningful buildings for a bigger goal, presented as imperative in order to move back to the future.

The ongoing ruination of Kino Rodina has happened in this kind of context; hence the building is not simply a symbol of the abnormality of socialist materialities and their supposed denial of the effects of time (Fehérváry, 2013), but also a signifier of official indifference, negligence and abandonment (Dzenovska, 2020; Martínez, 2022). In their study of the heritage-making discussions about Sillamäe, Saara Mildeberg and Jaanika Vider (2022) point out that a 'silent revenge' (a sort of passive aggressive strategy) has been the hegemonic approach towards the Soviet-era material culture. They foreground that in Estonia, Soviet heritage is institutionally considered of displaying an undesirable and dissonant kind, because 'the physical remains of the past offer up an identity that many of those in the present wish to distance themselves from, even while at the same time recognizing it fully as a part of their history' (ibid.: 2). The situation in Sillamäe, at least as I experienced, is slightly different than in the rest of the country, as for many locals the Soviet past and its remains still operates as a meaning-making apparatus, despite being treated as wasteful in hegemonic institutional representations and official management.

Above, I mentioned the headquarters of the Sillamäe Herald. The heating system therein has been stolen, also the lamps. So, we cannot feel this building is inhabitable. Indeed, it rather appears as beyond repair. When strolling around, you can detect traces of previous trespassers, such as graffiti on the walls and excrement on the floor. Someone seems to have used the darkness of this corner of the world to do his basic, scatological needs indeed. The feather on the stairs might also be a testimony of how the building has been transformed into a pigeon loft. I keep walking and see a series of toy guns laying on the floor, which indicates that the building has become a playground for some. There is a wide range of graffiti on the walls – from CR7 football player and Red Bull logos to rather existentialist messages and drawings. There are also matchboxes, broken glass, peeled wallpaper, headphones, books, banners from the social democrat party and random newspaper sheets from different years laying on the floor, which itself is full of various obstacles. Then, I found leftover shoes. And rotten animals. They both unsettle me. How shall I interpret this?

In addition to this, and in the context of the Russian invasion of Ukraine, the Estonian state invited the municipality of Sillamäe to rename some of its streets – including the Rumyantsevi tn., named after Mihhail Rumyantsev, a resident of Sillamäe who died in the war in Afghanistan. The street was already renamed in 1989, replacing the one named after Andrei Zhdanov, a Soviet politician and cultural repressor during the years of Stalin. However, the municipal council rejected the invitation, holding that Rumjantsev 'has not

Figure 9.9 Former headquarters of the Sillamäe Herald (Photo: Francisco Martínez, 2023).

Figure 9.10 Former headquarters of the Sillamäe Herald (Photo: Francisco Martínez, 2023).

committed any illegal acts or fought against the Republic of Estonia',[4] the argument that was seemingly convincing for regional minister Madis Kallas, who agreed in keeping that name.

Architectural taxidermy

As noted in the introduction to this volume, heritage is not given, but is a field of action. Heritage is, first and foremost, the product of a valuing and selecting process in which a particular idea of the content and contour of a community is enacted. Thus, around heritage-making there is an expectation for an object to contribute to the definition of communities, giving form and content to identities. Hence, heritage is often acknowledged as an affirmative recognition and invoked as having a positive quality (see also Harrison, 2013).

Different kinds of practices of value allocation and re-presentation take place in the categorization of something as heritage – documenting, archiving, conserving, communicating, insuring, etc. For a heritage object to be *made* – imposing certainty over ambivalence – a complex set of socio-technical practices are mobilized towards securing the stability of the heritage object in turn, including cultural and legal appropriation. In some cases, it might begin by unexpected encounters with the weird and ambivalent, alongside the gesture of noticing, as pointed out by Hilary Orange in the conclusion of this volume.

Based on arguments around historical representation, political symbolism, touristic potential and financial profit, the heritage object is materially fixed, culturally frozen and ecologically stable, so the different elements composing it do not move (Domínguez Rubio, 2020). Alas, this does not mean that heritage meanings cannot be contested or generated anew. In some cases, problematic legacies have already been recognized as meaningful heritage, even if having awkward connotations and despite opening up social divisions and dissonance instead (Tunbridge and Ashworth, 1996; Macdonald, 2009). Therefore, it is important to consider how ambivalent heritage can be reintegrated in the present, overcoming the supposed contradiction between preserving and using. Alas, in countries such as Estonia, the demolition of postindustrial remnants has been largely understood as a post-mortem act of justice, since industrial modernization was widely brought here by the Soviet regime (Martínez and Agu, 2021). And yet, demolition is always costly, produces multiple kinds of waste and tends to deploy arrogant, delusional anthropocentric acts of domination.

In *Remains of the Soviet Past in Estonia* (Martínez, 2018a), I described a series of buildings in Tallinn that faced demolition or went through taxidermy interventions, even if they were well-functioning. As I pointed out, the pastness attributed to buildings came not as an intrinsic quality but as a contested practice. In the attempt to construct a new tradition, several Soviet landmarks were sacrificed or dissected in the Estonian capital. For instance, we saw impressive buildings torn down to make space for shopping malls (i.e. Sakala keskus). Other buildings were privatized and suffered an experiment of taxidermy (Postimaja, Teenindusmaja, etc.), or linger in an unclear status (i.e. Linnahall). They were not dead but enduring, despite receiving limited investment and maintenance work.

The case of Kino Rodina differs, however. First, because the building has not had any use or purpose for decades; second, because it is located in a town characterized by postindustrial dynamics and outmigration; and third, because the ownership lays in private hands and the owners are incapable of actively preserving the building. As a result, the standing of this building is ambivalent, as it is simultaneously dead and alive, and with an unclear fitting in the present. Somehow, it reminds me of a zombie, still standing and holding the organs, but with a 'lifeless' existence and creating disorder in the surroundings. The zombie metaphor is associated with ambivalence, while the one of taxidermy is related to fixing and the certainty of forms. In any case, taxidermy or zombiism, these forms of heritage recognition without preservation can hardly be considered as neutral or apolitical, but rather point at different correlations of forces and weaknesses.

In taxidermic renovation of a heritage building, only the carcass is kept to suit the new commercial use, taking away the warmth of previous occupations, thus sending the message to the social world that these buildings used to represent oblivion and turning these buildings into mummies, without their internal organs left.[5] Some of the heritage listed buildings in Tallinn, such as Postimaja or Patarei, saw their interior cleared out and removed. But that has not yet happened in the Kino Rodina of Sillamäe.

In this case, the building deteriorates because of exposure to seasonal changes, because of its own material affordances and design errors and, overall, because of a lack of care – even if in heritage sites, objects are systematically categorized and formally fixed and predictable. Taxidermy or zombiism are both forms of heritage recognition without preservation. Besides sending the social world to oblivion, an important feature of the architectural taxidermy is that buildings were often not dead before the socio-technical intervention, but rather well-functioning. In the case of zombie heritage, the buildings appear presently dead, but the organs and intestines of the past social world remain warm.

As a critique of too-strict, modern strategies of heritage preservation, new approaches have emerged claiming that *let it be* strategies can be generative in themselves. An example of this is the proposal by geographer Caitlin DeSilvey (2017: 326) for an 'arrested decay … to uphold their structural integrity yet preserve their ruined appearance'. This take on heritage preservation, often related to postindustrial sites, proposes a lack of intervention or remediation as a form of landscape curation, as 'a strategy of low-impact conservation aimed at maintaining architectural ruins in a fragile stasis' (Moshenska, 2015: 78). One complication, however, is that institutions tend to show concern for uins – as heritage – but being rather intolerant for ruination – as that is considered a rather messy process outside of official heritage discourses (Pétursdóttir and Olsen, 2014). Another complication is that important traces might disappear themselves when at the mercy of social changes and weather conditions.

The Soviet Union, as a political regime, has long ended. But as a material culture it is still vanishing, rather quick and merciless nowadays. The disintegration of the USSR itself led to an exercise of intensive undoing through the unmaking of both material legacies and the collective memory of the past regime. Remains of the recent past are rapidly disappearing in Estonia, turning the study of post-socialism into an exercise of accelerated archaeology. Perhaps that exercise is not just accelerated, but also hopeless, once assuming that it might last decades for something contemporary to become an archaeological object of study, as suggested by Alfredo González-Ruibal (2019).

The disappearance and abandonment of Soviet material culture is making it difficult to understand the recent past, hence the need to reflect on the heuristic relevance of these traces. Consequently, we can question whether what is actually vanishing is the possibility of learning from the past, since it is the physicality of things that allows future generations to engage with remains (gaining insights into the societies that produce them as they represented themselves). Such an archaeological gesture invites us to think how buildings might persist in their remaining, since their durability makes them liable to represent unexpected attributes in the present. But what remained from that era is often going through a social and institutional disinvestment. Indeed, what makes buildings such as Kino Rodina an ambivalent heritage is that these objects do not disappear in a neat way and remain questioning traditional understandings of heritage as a cultural closure and the termination of a socio-material process.

Conclusion

This chapter considered the ambivalent standing of postindustrial ruins by contraposing the concepts of 'architectural taxidermy' and 'zombie heritage' to understand the situated relationship between material and social change in heritage making. The reader was welcomed to Sillamäe, the first area where a Soviet built form would be under protection in Estonia. The study puts the focus on the Kino Rodina, a movie theatre recognized as heritage but remaining without preservation. Kino Rodina thus is a meaningful example of ambivalence, since it has failed to disappear, nor be systematically preserved as a heritage object. Nowadays, the abandoned building calls for public attention, pending to be repaired in a wider context of negative capability.

Even in an ambivalent standing, zombie heritage can still play an active role in the production and sustenance of relations and culture, endowed with particular forms of value and magnetic power. However, these meanings and ideas might be unwelcome or integrated in the present with difficulty. Such is the case with Soviet legacies in Estonia, where the socialist past has institutionally been presented as radically separated from the present and unfitting. Kino Rodina stands in the middle of it, in the private hands of owners that have no means nor will to maintain it, and in a nation-state that reconstructed its legitimacy in opposition to what this form of heritage politically represents. In the meantime, the building falls apart and the possibility of learning from the past diminishes.

While revising the proofs of this chapter, I learnt that both buildings have been bought by an entrepreneur from Tartu; the former headquarters of the Sillamäe Herald to be transformed into a hotel spa and Kino Rodina "as a present to his wife".[6] Perhaps Kino Rodina has been preserved by its dead standing and lack of use, and not vice versa. Perhaps the zombie is us, and not the building.[7]

Notes

1 As recognized by officials of the Heritage Board in conversations held privately, as well as covered by the national media. See for instance a recent note by the Estonian News Agency: https://news.err.ee/1608606160/heart-of-sillamae-buildings-to-be-placed-under-heritage-conservation (retrieved August 2023).

2 That version of history, as if nothing was here before industrialization, is however contested by ethnic Estonians, as for instance in the military museum of Vaivara.

3 Aleksandr asks me for help to sell the building, so if any of the kind readers of this volume are interested in an old Soviet movie theatre, please send me an email and I will put you both in touch: francisco.martinez@tuni.fi.
4 See coverage in the Estonian News Service, https://news.err.ee/1609020197/ministry-rejects-request-to-change-sillamae-street-name (retrieved August 2023).
5 The earliest known taxidermists were the ancient Egyptians. But, the term itself derives from the Greek word *taxi*, meaning 'movement', and *derma*, meaning 'skin'. The process consists of stretching skins over wooden or foam forms to reproduce how animals looked like and mimic their postures and gestures simulacra-like. In Britain, taxidermy was especially popular in the Victorian era, and first practiced by explorers to preserve creatures and later used by aristocrats to impress their guests. Taxidermy has been regaining interest from post-humanist approaches because of the capacity of this practice to engender political realities (see Marbury, 2014; Aloi, 2018), and also because of its potential for touristification and real-estate speculation (Martínez, 2018b). A possible alternative would have included chimeras – entities composed of parts from different tangible bodies.
6 I heard the news from a local real estate entrepreneur during my visit to Sillamäe on 18 May 2024.
7 This work was financially supported by the European Union [ERC, WasteMatters, grant number: 101043572] and the Academy of Finland [grant number: 350191]. Views and opinions expressed are those of the author only and do not necessarily reflect those of the EU, the European Research Council Executive Agency or any other funder named.

Bibliography

Aloi, G. (2018), *Speculative Taxidermy: Natural History, Animal Surfaces, and Art in the Anthropocene*, New York: Columbia University Press.

Bauman, Z. (1991), *Modernity and Ambivalence*, Cambridge: Polity.

Buchli, V. (1999), *An Archaeology of Socialism*, Oxford: Berg.

Buchli, V. (2013), *An Anthropology of Architecture*, London: Bloomsbury.

Cairns, S. and J. Jacobs (2014), *Buildings Must Die*, Cambridge, MA: MIT Press.

DeSilvey, C. (2006), 'Observed Decay: Telling Stories with Mutable Things'. *Journal of Material Culture* 11 (3): 318–88.

Domínguez Rubio, F. (2020), *Still Life: Ecologies of the Modern Imagination at the Art Museum*, Chicago: Chicago University Press.

Dzenovska, D. (2020), 'Emptiness and Order', *Cultural Anthropology Forum* (retrieved June 2023) https://culanth.org/fieldsights/emptiness-and-order.

Fehérváry, K. (2013), *Politics in Color and Concrete*, Bloomington: Indiana University Press.

González-Ruibal, A. (2019), *An Archaeology of the Contemporary Era*, London: Routledge.
Gordillo, G. (2014), *Rubble: The Afterlife of Destruction*, Durham, NC: Duke University Press.
Harris, N. (1999), *Building Lives: Constructing Rites and Passages*, New Haven: Yale University Press.
Harrison, R. (2013), *Heritage: Critical Approaches*, Abingdon: Routledge.
Lynch, K. (1990), *Wasting Away*, San Francisco: Sierra Club.
Macdonald, S. (2009), *Difficult Heritage: Negotiating the Nazi Past in Nuremberg and Beyond*, London: Routledge.
Marbury, R. (2014), *Taxidermy Art: A Rogue's Guide to the Work, the Culture, and How to Do It Yourself*, New York: Artisan.
Martínez, F. (2018a), *Remains of the Soviet Past in Estonia: An Anthropology of Forgetting, Repair and Urban Traces*, London: UCL Press.
Martínez, F. (2018b), '"This Place has Potential": Trash, Culture, and Urban Regeneration in Tallinn, Estonia', *Suomen Antropologi*, 42 (3): 4–22.
Martínez, F. (2022), 'Memory, Don't Speak! Monumental neglect and memorial sacrifice in contemporary Estonia', *Cultural Geographies*, 29 (1): 63–81.
Martínez, F. (2024), 'Afterword: The Only Real Nation is Ruination' in *Haunting Ruins*, C. Calzana and V. Gamberi (eds), Oxford: Berghahn.
Martínez, F. and M. Agu (2021), 'Postcards from the Edge: Territorial Sacrifice and Care in Eastern Estonia', *Roadsides*, 5: 1–7.
Mildeberg, S. and J. Vider (2022), 'Soviet Heritage(scape) in Sillamäe: Documenting the Potential in an Emerging Tourism Destination', *MDPI Societies*, 12 (5): 1–19.
Moshenska, G. (2015), 'Curated Ruins and the Endurance of Conflict Heritage', *Conservation and Management of Archaeological Sites*, 17 (1): 77–90.
Orlova, A. [Орлова, А.] (2005), 'Kino Rodina' [Кинотеатр "Родина"], in *Sillamäe 1946–1989* [*Силламяэ (1946–1989 г.г.)*], Self-published in Sillamäe, pp. 135–7.
Pétursdóttir, Þ., and B. Olsen (eds) (2014), *Ruin Memories*, London: Routledge.
Pusca, A. (2010), 'Industrial and Human Ruins of Postcommunist Europe', *Space and Culture*, 13 (3): 239–55.
Storm, A. (2014), *Post-Industrial Landscape Scars*, New York: Palgrave Macmillan.
Tunbridge, J. E. and G. J. Ashworth (1996), *Dissonant Heritage: The Management of the Past as a Resource in Conflict*, London: Wiley.
Wilson, J. Q. and G. L. Kelling (1982), 'The police and neighborhood safety: Broken windows', *The Atlantic Monthly*, 249 (3): 29–38.
Wirth, L. (1938), 'Urbanism as a Way of Life', *American Journal of Sociology*, 44: 1–14.

10

Discussion: The Discomforting Appeal of Ambivalent Heritage

Hilary Orange

This edited volume is a welcome and timely reminder to consider ambivalence – in others, in oneself, in heritage practice and in academia. According to Bauman (1991), ambivalence is the unavoidable outcome of the ordering obsession of modernity; one that is always ill-fated as the process of ordering only serves to stimulate ambivalence, here conceived as a type of waste product (1991: 1–3). Bauman's 1991 volume *Modernity and Ambivalence* was written in Leeds at a time of UK-wide deindustrialization and the burgeoning of a Conservative-government-sponsored, nostalgia-driven heritage industry. Neither heritage nor industry are discussed in the volume directly, but his opening critique, on ambivalence as the alter ego of linguistic order, still provides ample room to reflect on the way that the classificatory and managerial styles of heritage practice produce systems of care and protection that are synonymous with ordering the 'olden'. His discussion on modern intellectualism's quest to 'exterminate' ambivalence (1991: 7) resonates with the desire to produce academic writing that provides a strong position. If ambivalence and order are co-existent and sustain each other, this discussion chapter owes a debt to various states and stages of ambiguity.

Thus, following Bauman, heritage and academia necessarily feed off and create ambivalence as part of their projects, but *ambivalent heritage* is strangely a term that has yet to gain common traction in heritage discourse, despite publications pre-dating this volume, for instance Chadha's work on the ambivalent nature of colonial cemeteries in Calcutta (2006). I hope that this volume brings attention to the ambiguous things and feelings in heritage contexts that do not quite fit and are not just one thing but are again another. In considering various ambivalent encounters, the authors in this volume connect bodily, cognitive and sensorial experiences with entanglements of liminality and imagination in ways that are open-ended and, at times, may question reality. The archaeofantastical 'Temple of Lemminkäinen' is one example in this volume of things that may 'appear to be one thing but then turn out to be something else' (Herva et al., Chapter 6 this volume, p. 149). Thus, certainty and matter may become soluble. I imagine words in print that dissolve, reform and become something else.

On ambiguity, an image comes to mind. *The Storyteller* (1986), by the Canadian artist Jeff Wall, depicts three groups of sitters who are occupying a nondescript grassy embankment close to the shelter of a nondescript bridge that passes over a presumably nondescript and out-of-shot highway. One group of sitters, around the remnants of a fire, reference the seated figures in *Le Déjeuner sur l'herbe* by Édouard Manet (Figure 10.1), a painting that shocked Parisian society in the 1860s by subverting the fine art traditions of the bucolic, the picturesque and of portraiture. In the painting, a nude woman, possibly a sex worker, sits with two fully clothed men. In *The Storyteller* the figures look at ease on the embankment, a space designed to support transience and travel, and they

Figure 10.1 *Le Déjeuner sur l'herbe* by Édouard Manet (1862–3). Public Domain, https://commons.wikimedia.org/w/index.php?curid=21855901.

counter the idea of such places as 'non-places', rather than meeting places (Augé, 1995). The Metropolitan Museum of Art (2023) holds the photograph in its collection and interprets it as being 'ultimately hopeful, holding in suspension the potential for cultural traditions to survive and contest historical amnesia', but the liminal setting creates a deeply ambivalent image of people seemingly out-of-place and potentially unsafe.

Over time, I have returned to Wall's image *because* of its ambivalent pull. In Chapter 8, Äikäs et al. point out the positive associations that ambivalent encounters with heritage may bring and, in this case, the intellectual uncertainty that *The Storyteller* conjures appeals as the figures' back-stories are unknowable. I wonder where they have come from, how long they stay and of the push-pull dynamics that led them to this place (of course, Wall's photograph is an elaborate, staged tableau but still it feels like it could be real). At other times, ambivalent heritage can be deeply discomforting and problematic. A memory comes to mind.

It is July 2016 and I am in Camden Square in London, standing at the edge of the park at the centre of the Square directly opposite the former home of singer/songwriter Amy Winehouse, who died in 2011. I am there to record the tributes left by fans to celebrity musicians as part of a research project on rock and pop memorials looking at shrines and tributes to David Bowie, George Michael, Mark Bolan, Freddie Mercury and Winehouse (Graves-Brown and Orange, 2017; Orange and Graves-Brown, 2019). In the case of Amy Winehouse, a makeshift shrine developed around the trunks of three London plane trees that overlook her former home. At some point, Camden Council wrapped bamboo matting around the trunks of the trees, presumably to prevent the bark being damaged by graffiti or taken as souvenirs, and the matting provides small nesting spaces for notes from fans, for flowers, business cards and other tribute paraphernalia. I had previously recorded empty cigarette packets and cigarette stubs pushed into the matting as well as empty alcohol bottles left at the base of the trees (Figure 10.2). I had visited this spot several times before, but on this occasion, I fell into conversation with a man walking his dog who told me that he had found and removed hypodermic needles stuck into the trees. Amy was a smoker. She died of alcohol poisoning and had a well-documented drug habit, something that the British tabloid press were often keen to remind their readers of. A mile away a statue of Winehouse was erected in 2014 in Camden Market, a tourist hot spot. The statue is inherently tactile. Placed at ground height, with her arm on her hip, statue-Amy waits for passing visitors to link arms with her or drape their arms around her shoulders for selfies and photographs (Figure 10.3).

Figure 10.2 Shrine to Amy Winehouse, Camden Square Gardens, London (Photo by author).

Figure 10.3 Statue erected in honour of Amy Winehouse in 2014 in Camden Market (Photo by author).

The act of paying tribute, through celebrity veneration, combines simultaneous grief and pleasure; the pleasure comes in choosing or making a tribute object and in undertaking a pilgrimage to a destination in a way that creates a deeper physical, emotional or spiritual connection with the object of veneration (Vikan, 1994; Margry, 2008: 17). It is not uncommon for fans to select items that have a biographical relationship to the life of the artist. However, when compared with other tribute sites to celebrity musicians in London, which are typified by album covers, flowers, teddy bears, references to lyrics and quasi-spiritual messaging (Orange and Graves-Brown, 2019), a characteristic of the Winehouse memorials is her body: what she did with it, what she put in it and the passive work that a proxy of her body does to construct transient tourist experiences. Although acknowledging that music pilgrimage is a democratic form of self-expression that raises 'fan cultural capital' (Hills, 2002) and can be a form of healing (Recuber, 2012), I find the manner in which the tribute objects highlight her female body and potentially glamourize the manner of her death, rather than her music, troubling.

The needle stuck in the tree sticks in my mind as an object out of place, but one that is nevertheless doing the messy work of occupation. By occupation, I mean the occupation of space and being occupied with *doing* something, e.g. a pursuit, a hobby or a role within a cultural ecosystem. It is notable that several of the chapters in this volume are concerned with care, responsibility and the transient occupation of abandoned industrial sites and other off-the-beaten-track locations.

Industrial spaces have long appealed to children and adults alike for their seclusion. In their wild(er) potential for creative modes of expression, all kinds of things can happen away from the constraints of public gaze and supervision (Edensor, 2005; Orange, 2019). In Chapter 8 (Äikäs et al.), the hobby of geocaching, for example, responds to a perceived loss of freedom in hyper-ordered museum spaces. Those who (re)visit, including the researchers themselves, reanimate these industrial spaces (Orange, 2015) and in Martínez's (Chapter 9, p. 216) terms places are warm when occupied and used. A substantial part of the attraction of these out-of-the-way locales is their unusual aesthetics, and the opportunity for heightened sensory engagements with the more-than-human. Sound, smell and touch thus run through the volume. In the abandoned buildings of the secret, nuclear town of Sillamäe, peeling-off ceilings overlook invoices on the floor, and dirty curtains flap in front of broken windows (Martínez, Chapter 9). The smell of pine forests accompanies tuberculosis patients seeking privacy and pleasure outside of the walls of their sanitorium

(Mäki and Ruotsala, Chapter 3), and search and rescue dogs sniff around derelict industrial areas during training exercises (Chapter 8, Äikäs et al.). On the sense of sight, people move differently through different qualities of light and dark, and some of the chapters evoke a cinematic Lynchian shadow-world. In Chapter 8, Äikäs et al. consider how industrial sites, such as factory halls, cellars and other (semi-)abandoned structures, create dissonance for hobbyists and other users (such as search and rescue dog trainers). Here, dark corners create a sense of adventure, and one urban explorer describes getting used to 'the smells of mould, dirt and grease, of old houses or cellars'. I wonder about the extent to which familiarity dissipates ambivalent feelings towards a place.

Spiwak provides a good example of the things that can happen away from public sight. In a slate mine in North West Wales, a collection of small exhibits are curated by mine explorers along shelves and slabs, evoking a DIY Punk ethos (Morgan, 2015) that is 'uninfluenced by "expert" narratives' (Chapter 1, p. 24). On this Spiwak (p. 19) writes that 'communities of explorers, caretakers and vandals have made these spaces their own'. This is an apt reminder that away from the codes of surface life 'some choose to be caretakers and storytellers' (p. 36). I am reminded of a garden in a coal mine. In the 1920s, in Nova Scotia, a German miner called William Krause created Canada's first underground garden in 1B coal mine, Glace Bay, Cape Breton. The garden was originally lit by electric light and fertilized with horse manure (pers. comm. Cape Breton Miners Museum). Mining ended in the 1980s and a museum now runs mine tours that stop at a replica of the garden, albeit now planted with plastic flowers. At the time of my visit in June 2023, this seemed an oddity, but on reflection there is an equivalence with bringing a potted plant into an office workspace. In both of these cases, the mine garden and the underground collection, there is a synergy with *The Storyteller*, in claiming personal space within an interstitial buffer zone that represents mental and physical detachment from everyday/'normal' life (Abdel and Pintos, 2022). As Spiwak (Chapter 1, p. 18) explains, the slate mines may be chaotic spaces, but they are inscribed on the UNESCO World Heritage List, and in them 'decay and curation co-exist'. The display of objects underground points to networks of human and non-human actors involved in forming and reforming archaeological objects in ways that mirror above ground collecting processes (Bennett et al. 2017), albeit in damp and gloomy conditions.

Like the sanitorium in Paimio, Finland (Chapter 3), plans have also been mooted to include the Chernobyl zone in the UNESCO World Heritage Site list. In Chapter 5, Ojala discusses dark heritage and the emergence of the Chernobyl zone as a destination for tourism, and reveals how tourism works in tandem with multitextured heritage practices, where 'The risks of radiation were also acknowledged but bargained for between the acquisition of "real" and personal experience' (p. 120). In estimating risk to oneself and others, Ojala (p. 110) writes that:

> The city of Pripyat, the Soviet utopia Atomgrad, and its contemporary ruins, are primary examples of these types of ambivalent heritage resources, which are interpreted through individual meaning-making processes and spatial strategies, and evoke strong sensory experiences, such as nostalgia, sublime, pain and discomfort, in addition to the joy and mystery of the atmospheres of abandoned ruins.

In the above quote, Ojala makes the important point that sites with repellent and ambivalent histories can have a magnetic attraction, not least through their appeal to the imagination, to novelty, for the heightened adrenaline rush they bring and for the personal challenge of risk-taking. These facets of ambivalent heritage perhaps explain why so many of the chapters in this volume have chosen to focus on abandoned industrial sites. Large-scale industrial remains, whether above or below ground, are typically viewed through intertwined demonic, romantic and heroic lenses (Alfrey and Putnam, 1992: 40), and the combination of terror and romance that characterizes the industrial sublime has a deep, emotional power. Abandoned industrial locations often appear as key locations in horror movies, and as Godin (2022) has explored, monstrosity, archaeology and the detritus of the post-human world intersect closely with the genre of horror. In the survival monster horror, for example, *A Quiet Place Part II* (2020), the Abbot family hide in the former Bethlehem Steel Lackawanna plant. The steel plant is both a refuge and a death trap. In one scene the Abbot's son is trapped in a furnace. Interestingly, the site also serves as the backdrop for one of the movie's release posters. In other examples, a pivotal early scene in *It Follows* (2014) was shot in the empty halls of the abandoned Packard Plant in Detroit, and some scenes of *The Silence of the Lambs* (1991) were filmed at the abandoned Westinghouse turbine factory. There are many other examples from the horror, thriller and music video genres. What is clear is that abandoned industrial locations thrill, while active industrial sites, humanized by the routines of labour, do not carry the same kind of punch.

There is a difference, however, between the thrill seekers, the cinema goers, the tourists and the urban explorers who welcome, or at least tolerate, ambivalent experiences through choice, and former workers whose memories of work will be rekindled, rather than imagined as forms of post-memory. Shelley used Frankenstein's monster (1818) to represent the anxieties that accompanied rapid technological change during the Industrial Revolution (Godin, 2022: 120), and the years between then and now have been patterned by labour exploitation, gender, class, race and union struggles, environmental degradation, accidents, mortality, forced labour and disease. Transient exposure provides a thrill, chronic exposure could be life threatening. Visiting is discretionary, working is not if you have got yourself or a family to support. The context of a miner, who needs to carry on working, but struggles to breathe because of emphysema, or any other industrial disease (McIvor, 2023), is markedly different again. A comparison could be with tourists versus former servicemen visiting military sites.

I return again to Wall's *The Storyteller* and the unanswered questions I have about the image. Does it depict hope or homelessness, home or the push-pull of migration across borders, land possession or dispossession, a place to care and curate identity and tradition, or something else – something else that is monstrous? The difference while asking these questions is that I am safe in my questioning, rather than sheltering on an embankment. Matila (Chapter 4, pp. 92–3) usefully questions an 'ambivalence towards the industrial past and a conflict with what it gave people and what it took', and I wonder how ambivalence fares in the context of industrial work when participation in labour systems is typified by a lack of autonomy and where abandoned space is replaced by the routines of labour.

The largest steel plant in Wales is located in the town of Port Talbot in South Wales. In 2018, the World Health Organization wrongly labelled Port Talbot as one of the most polluted areas in the UK, for which the organization later apologized, but the title has stuck. Just before Christmas in 2018, Banksy chose a nondescript garage in the town to create a new graffiti artwork called *Seasons Greetings* to highlight the issue of air pollution (Figure 10.4). On one side of the garage, a child plays happily in falling snow, on the other side the snow turns out to be ash falling from a dumpster fire. An article in *The Independent* in December 2018 starts 'Dear Banksy: Port Talbot doesn't deserve to be an emblem of the global pollution crisis' before pointing out the 'bittersweet reality of life in industrial Wales'. That industry equals pay (Lewis, 2018). A *Guardian* article from 2023 refers to the plant as the 'economic heart of the area … employing 4,000 people and an estimated 10,000 in the wider community' (Jolly, 2023). Photographs of old Port Talbot line the town's Wetherspoon's pub, and show what the town looked like before an elevated section of the M4 motorway was built through the town in the 1960s, and before the old town centre was demolished for the building of a new indoor shopping centre in the 1970s. The town has no museum, but a large Facebook group 'Port Talbot Old and New' has

Figure 10.4 *Season's Greetings* by Banksy. Banksy, CC BY-SA 4.0 <https://creativecommons.org/licenses/by-sa/4.0>, via Wikimedia Commons.

25,000 members who story-tell in the absence of any 'official' interpretation. As an outsider, I am intrigued about the town and how the ambivalent feelings that Matila describes, born out of an acknowledgement of a pride in the history of a steel town, as well as a history of industrial accidents, environmental degradation and climate crisis, will develop within the curation of Port Talbot's digital memory. Any ambivalent feelings are likely to be complex and deeply personal in being connected to place, memory, emotional and sensory states, affordance and loss.

To conclude, why might we like to know about ambivalent heritage? At a time when media and populism (to the left and right of the spectrum) are driving partisan and polarized thinking, ambivalent heritage represents the grey zone, but one that is full of nuance and colour. Consider what the opposite of ambivalent heritage might be. The Authorized Heritage Discourse as discussed by Smith (2006: 11), sets out the problematic institutionalization of hegemonic, Western elite cultural values by state actors. In a similar vein, a 'balanced' heritage would constitute complete epistemological and ontological assurance in ways that would be banal and infeasible given the attendant biases, inequalities, power plays and questionable ethics that accompany heritage making. Balanced heritage, at its extreme, would be the stuff of totalitarian dreams.

What the authors collectively point to in this volume is finding ways to give mental and physical attention to complexity and, in so doing, acknowledging ambivalence as an innate condition of being human amidst the more-than-human wonder, building, waste and detritus of the world. As Äikäs et al. (Chapter 8, p. 195) say, 'what one might see as a site to be demolished, another can see as a place of adventures'. We do not and should not see the world in the same way and therein is hope.

References

Abdel, H. and P. Pintos (2022), 'Interstitial Spaces and Public Life, the Overlooked Interventions that Weave our Built Environment', *ArchDaily*. Available online: https://www.archdaily.com/984818/interstitial-spaces-and-public-life-the-overlooked-interventions-that-weave-our-built-environment.

Alfrey, J. and T. Putnam (1992), *The Industrial Heritage: Managing Resources and Uses*, London: Routledge.

Augé, M. (1995), *Non-Places: Introduction to an Anthropology of Supermodernity*, London: Verso.

Bauman, Z. (1991), *Modernity and Ambivalence*, Cambridge: Polity.

Bennett, T., F. Cameron, N. Dias, B. Dibley, R. Harrison, I. Jacknis and C. McCarthy (2017), *Collecting, Ordering, Governing: Anthropology, Museums, and Liberal Government*, Durham, NC: Duke University Press.

Chadha, A. (2006), 'Ambivalent Heritage: Between Affect and Ideology in a Colonial Cemetery', *Journal of Material Culture*, 11 (3): 339–63. https://doi.org/10.1177/1359183506068809.

Edensor, T. (2005), *Industrial Ruins: Spaces, Aesthetics and Materiality*, New York: Berg.

Godin, G. (2022), 'Monstrous things: horror, othering, and the Anthropocene', *Post-Medieval Archaeology*, 56 (2): 116–26.

Graves-Brown, P. and H. Orange (2017), '"The Stars Look Very Different Today": Celebrity Veneration, Grassroot Memorials and the Apotheosis of David Bowie', *Material Religion*, 13 (1): 121–3.

Hills, M. (2002), *Fan Cultures*, London: Routledge.

It Follows (2014), [Film] Dir. David Robert Mitchell, USA: Radius-TWC, The Weinstein Company.

Jolly, J. (2023), 'Welsh steelworks at breaking point over the cost of cleaning up its act', *The Guardian*, 15 July 2023. Available online: https://www.theguardian.com/business/2023/jul/15/welsh-steelworks-at-breaking-point-over-the-cost-of-cleaning-up-its-act.

Lewis, S. (2018), 'Dear Banksy: Port Talbot doesn't deserve to be an emblem of the global pollution crisis', *The Independent*, 20 December 2018. Available online: https://www.independent.co.uk/voices/banksy-port-talbot-graffiti-wales-michael-sheen-steel-pollution-environment-a8692821.html.

Margry, P. J. (2008), *Shrines and Pilgrimage in the Modern World: New Itineraries into the Sacred*, Amsterdam: Amsterdam University Press.

McIvor, A. (2023), *Jobs and Bodies: An Oral History of Health and Safety in Britain*, London: Bloomsbury.

Morgan, C. (2015), 'Punk, DIY, and Anarchy in Archaeological Thought and Practice', *Online Journal of Public Archaeology*, 5: 123–46.

Orange, H. and P. Graves-Brown (2019), '"My death waits there among the flowers." Popular music shrines in London as memory and remembrance', in S. De Nardi, H. Orange, S. High and E. Koskinen-Koivisto (eds), *The Routledge Handbook of Memory and Place*, 345–56, London: Routledge.

Orange, H. (2015), *Reanimating Industrial Spaces: Conducting Memory Work in Post-Industrial Societies*, London: Routledge.

Orange, H. (2019), 'Uneven Surfaces: Bodily Engagements with the Postindustrial Wild', in George S. Jaramillo and Juliane Tomann (Eds), *Transcending the Nostalgic: Landscapes of Postindustrial Europe beyond Representation*, 43–63, New York: Berghahn.

Orange, H. (2022), 'Uneven Surfaces: Bodily Engagements with the Postindustrial Wild', in G. S. Jaramillo and J. Tomann (eds), *Transcending the Nostalgic: Landscapes of Postindustrial Europe beyond Representation*, 43–63.

A Quiet Place Part II (2020), [Film] Dir. John Krasinski, Paramount Pictures.

Recuber, T. (2012), 'The prosumption of commemoration: Disasters, digital memory banks, and online collective memory', *American Behavioral Scientist*, 56 (4): 531–49.

The Metropolitan Museum of Art, https://www.metmuseum.org/art/collection/search/286725.

The Silence of the Lambs (1991), [Film] Dir. Jonathan Demme, USA: Orion Pictures.

Shelley, M. (1818), *Frankenstein, or, The modern Prometheus*, London: Lackington, Hughes, Harding, Mavor, & Jones.

Smith, L. (2006), *Uses of Heritage*, London: Routledge.

Vikan, G. (1994), 'Graceland as locus sanctus', in G. DePaoli (ed.), *Elvis + Marilyn. 2X Immortal*, 150–70, New York: Rizzoli.

Index

Printed in the USA
CPSIA information can be obtained
at www.ICGtesting.com
LVHW021850221124
797368LV00003B/396
9781350426740